A BIOGRAPHY OF THE WORLD'S MOST FAMOUS EQUATION

E=mc²

by

David Bodanis

Publishing

First published in USA 2000
by Walker Publishing Co. and in Great Britain 2001
by Macmillan General Books

Published in Large Print 2001 by ISIS Publishing Ltd,
7 Centremead, Osney Mead, Oxford OX2 0ES
by arrangement with Walker Publishing Co. and
Macmillan General Books

British Library Cataloguing in Publication Data
Bodanis, David
　　E=mc2: a biography of the world's most famous equation.
　　– Large print ed.
　　1. Einstein, Albert, 1879-1955 – Influence 2. Space and time
　　3. Relativity 4. Large type books
　　I. Title
　　530.1'1

Library of Congress Cataloging-in-Publication Data
Bodanis, David
　　E=mc2: a biography of the world's most famous equation/
　　David Bodanis. – Large print ed.
　　1. Force and energy 2. Mass (physics) 3. Mathematical
　　physics 4. Einstein, Albert, 1879-1955 5. Large type books
　　– Specimens. I. Title

ISBN 0-7531-5606-7 (hb)
ISBN 0-7531-5607-5 (pb)

Printed and bound by Antony Rowe, Chippenham and Reading

Contents

Preface

A while ago I was reading an interview with the actress Cameron Diaz in *Premiere* magazine. At the end the interviewer asked her if there was anything she wanted to know. "Yeah," said Diaz, "What does E=mc^2 really mean?" They both laughed, then Diaz mumbled, "I mean it," and then the interview ended.

"You think she did mean it?" one of my friends asked, after I read it aloud. I shrugged, but everyone else in the room—architects, two programmers, and even one historian (my wife)—was adamant. They knew exactly what she intended: They wouldn't mind understanding what the famous equation meant too.

It got me thinking. Everyone knows that E=mc^2 is really important, but they usually don't know what it means, and that's frustrating, because the equation is so short that you'd think it would be understandable.

There are plenty of books that try to explain it, but who can honestly say they understand them? To most readers they contain just a mass of odd diagrams—those little trains or rocketships or flashlights that are utterly mystifying? Even firsthand instruction doesn't always help, as Chaim Weizmann found when he took a long Atlantic crossing with Einstein in 1921: "Einstein explained his theory to me every day," Weizmann said, "and soon I was fully convinced that he understood it."

I realized there could be a different approach. The overall surveys of relativity fail not because they're poorly written, but because they take on too much. Instead of writing yet another account of all of relativity, let alone another biography of Einstein—those are interesting topics, but have been done to death—I could simply write about E=mc². That's possible, for it's just one part of Einstein's wider work. To a large extent, it stands on its own.

The moment I started thinking this way, it became clear how to go ahead. Instead of using the rocketship-and-flashlight approach, I could write the biography of E=mc². Everyone knows that a biography entails stories of the ancestors, childhood, adolescence, and adulthood of your subject. It's the same with the equation.

The book begins, accordingly, with the history of each part of the equation—the symbols *E, m, c, =, and* ². For each of these—the equation's "ancestors"—I focus on a single person or research group whose work was especially important in creating our modern understanding of the terms.

Once the nature of those symbols is clear, it's time to turn to the equation's "birth." This is where Einstein enters the book: his life as a patent clerk in 1905; what he'd been reading, and what he'd been thinking about, which led to all those symbols he wove together in the equation hurtling into place in his mind.

If the equation and its operations had stayed solely in Einstein's hands, our book would simply have continued with Einstein's life after 1905. But pretty quickly after this great discovery his interests shifted to other topics; his personal story fades from the book, and instead we pick up with other physicists: more empirical ones now, such as the booming, rugby-playing Ernest Rutherford, and the quiet, ex-POW James Chadwick, who together helped reveal the detailed structures within the atom that could—in principle—be manipulated to allow the great power the equation spoke of to come out.

In any other century those theoretical discoveries might have taken a long time to be turned into practical reality, but the details of how Einstein's equation might be used became clear early in 1939, just as the twentieth century's greatest war was beginning. A long, central section of the book homes in on the equation's coming of age here, in the furious race between scientists based in the United States and those in Nazi Germany to see who could build a deathly, planet-controlling bomb first. The history is often presented as if America's victory were inevitable, due to the country's industrial superiority, but it turns out that Germany came dangerously closer to success than is often realized. Even as late as D Day in June 1944, Army Chief of Staff George Marshall saw to it that several of the U.S. units landing in France were supplied with Geiger counters as a precaution against a possible Nazi attack with radioactive weapons.

In the final sections of the book we switch away from war; the equation's "adulthood" has begun. We'll see how $E=mc^2$ probably is at the heart of many medical devices, such as the PET scanners used for finding tumors; it's also widespread in our ordinary household devices, including televisions and smoke alarms. But even more significant is how its power stretches far out into the universe, explaining how stars ignite, and our planet keeps warm; how black holes are created, and how our world will end. At the very end of the book, there are fairly detailed notes, for readers interested in more mathematical or historical depth. Additional background information is available at my Web site at dbodanis.com.

The stories along the way are as much about passion, love, and revenge as they are about cool scientific discovery. There will be Michael Faraday, a boy from a poor London family desperate for a mentor to lift him to a better life, and Emilie du Châtelet, a woman trapped in the wrong century, trying to carve out a space where she wouldn't be mocked for using her mind. Here are the stories of Knut Haukelid and a team of fellow young Norwegians, forced to attack their own countrymen to avert a greater Nazi evil; Cecilia Payne, an Englishwoman who finds her career destroyed after daring to glimpse the sun's fate in the year A.D 6 billion; and a nineteen-year-old Brahmin, Subrahmanyan Chandrasekhar, who discovers something even more fearful, out in the beating heat of the Arabian Sea in midsummer. Through all their stories—as well as highlights from those of Isaac Newton, Werner Heisenberg, and other researchers—the meaning of each part of the equation becomes clear.

Part One . . .

. . . Birth

CHAPTER ONE

Bern Patent Office, 1905

From THE COLLECTED PAPERS OF
ALBERT EINSTEIN, VOLUME I:

13 April 1901
Professor Wilhelm Ostwald
University of Leipzig
Leipzig, Germany

Esteemed Herr Professor!

Please forgive a father who is so bold as to turn to you, esteemed Herr Professor, in the interest of his son.

I shall start by telling you that my son Albert is 22 years old, that ... he feels profoundly unhappy with his present lack of position, and his idea that he has gone off the tracks with his career & is now out of touch gets more and more entrenched each day. In addition, he is oppressed by the thought that he is a burden on us, people of modest means....

I have taken the liberty of turning to you with the humble request to ... write him, if possible, a few words of encouragement, so that he might recover his joy in living and working.

If, in addition, you could secure him an Assistant's position for now or the next autumn, my gratitude would know no bounds....

I am also taking the liberty of mentioning that my son does not know anything about my unusual step.

I remain, highly esteemed Herr Professor, your devoted

Hermann Einstein

$E=mc^2$ · $E=mc^2$ · $E=mc^2$

No answer from Professor Ostwald was ever received.

The world of 1905 seems distant to us now, but there were many similarities to life today. European newspapers complained that there were too many American tourists, while Americans were complaining that there were too many immigrants. The older generation everywhere complained that the young were disrespectful, while politicians in Europe and America worried about the disturbing turbulence in Russia. There were newfangled "aerobics" classes; there was a trend-setting vegetarian society, and calls for sexual freedom (which were rebuffed by traditionalists standing for family values), and much else.

The year 1905 was also when Einstein wrote a series of papers that changed our view of the universe forever. On the surface, he seemed to have been leading a pleasant, quiet life until then. He had often been interested in physics puzzles as a child, and was now a recent university graduate, easygoing enough to have many friends. He had married a bright fellow student, Mileva, and was earning enough money from a civil service job in the patent office to spend his evenings and Sundays in pub visits, or long walks—above all, he had a great deal of time to think.

Although his father's letter hadn't succeeded, a friend of Einstein's from the university, Marcel Grossman, had pulled the right strings to get Einstein the patent job in 1902. Grossman's help was necessary not so much because Einstein's final university grades were unusually low—through cramming with the ever-useful Grossman's notes, Einstein had just managed to reach a 4.96 average out of a possible 6, which was almost average—but because one professor, furious at Einstein for telling jokes and cutting classes, had spitefully written unacceptable references. Teachers over the years had been irritated by his lack of obedience, most notably Einstein's high school Greek grammar teacher, Joseph Degenhart, the one who has achieved immortality in the history books through insisting that "nothing would ever become of you." Later, when told it would be best if he left the school, Degenhart had explained, "Your presence in the class destroys the respect of the students."

Outwardly Einstein appeared confident, and would joke with his friends about the way everyone in authority seemed to enjoy putting him down. The year before, in 1904, he had applied for a promotion from patent clerk third class to patent clerk second class. His supervisor, Dr. Haller, had rejected him, writing in an assessment that although Einstein had "displayed some quite good achievements," he would still have to wait "until he has become fully familiar with mechanical engineering."

In reality, though, the lack of success was becoming serious. Einstein and his wife had given away their first child, a daughter born before they were married, and were now trying to raise the second on a patent clerk's salary. Einstein was twenty-six. He couldn't even afford the money for part-time help to let his wife go back to her studies. Was he really as wise as his adoring younger sister, Maja, had told him?

He managed to get a few physics articles published, but they weren't especially impressive. He was always aiming for grand linkages—his very first paper, published back in 1901, had tried to show that the forces controlling the way liquid rises up in a drinking straw were similar, fundamentally, to Newton's laws of gravitation. But he could not quite manage to get these great linkages to work, and he got almost no response from other physicists. He wrote to his sister, wondering if he'd ever make it.

Even the hours he had to keep at the patent office worked against him. By the time he got off for the day, the one science library in Bern was usually closed. How would he have a chance if he couldn't even stay up to date with the latest findings? When he had a few free moments during the day, he would scribble on sheets he kept in one drawer of his desk—which he jokingly called his department of theoretical physics. But Haller kept a strict eye on him, and so the drawer stayed closed most of the time. Einstein was slipping behind, measurably, compared to the friends he'd made at the university. He talked with his wife about quitting Bern and trying to find a job teaching high school. But even that wasn't any guarantee: he had tried it before, only three years earlier, but never managed to get a permanent past.

And then, on what Einstein later remembered as a beautiful day in the spring of 1905, he met his best friend, Michele Besso ("I like him a great deal," Einstein wrote, "because of his sharp mind and his simplicity"), for one of their long strolls on the outskirts of the city. Often they just gossiped about life at the patent office, and music, but today Einstein was uneasy. In the past few months a great deal of what he'd been thinking about had started coming together, but there was still something Einstein felt he was very near to understanding but couldn't quite see. That night Einstein still couldn't put it together, but the next day he suddenly woke up, feeling "the greatest excitement."

It took just five or six weeks to write up a first draft of the article, filling thirty-eight pages. It was the start of his theory of relativity. He sent the article to *Annalen der Physik* to be published, but a few weeks later, he realized that he had left something out. A three-page supplement was soon delivered to the same physics journal. He admitted to another friend that he was a little unsure how accurate the supplement was: "The idea is amusing and enticing, but whether the Lord is laughing at it and has played a trick on me—that I cannot know." But in the text itself he began, confidently: "The results of an electrodynamic investigation recently published by me in this journal led to a very interesting conclusion, which will be derived here." And then, four paragraphs from the end of this supplement, he wrote it out.

$E = mc^2$ had arrived in the world.

Part Two . . .

. . . Ancestors of $E=mc^2$

CHAPTER TWO

'E' Is for 'Energy'

The word *energy* is surprisingly new, and can only be traced in its modern sense to the mid 1800s. It wasn't that people before then had not recognized that there were different powers around—the crackling of static electricity, or the billowing gust of a wind that snaps out a sail. It's just that they were thought of as unrelated things. There was no overarching notion of "Energy" within which all these diverse events could fit.

One man who took a central role in changing this was Michael Faraday, a very good apprentice bookbinder who had no interest, however, in spending his life binding books. As an escape hatch from poverty in London of the 1810s, though, it was a job that had one singular advantage: "There were plenty of books there," he mused years later to a friend, "and I read them." But it was fragmentary reading, and Faraday recognized that, just snatching glimpses of pages as they came in to be bound. Occasionally he had evenings alone, next to the candles or lamps, reading longer sixteen- or thirty-two-page bound sheaves.

He might have stayed a bookbinder, but when Faraday was twenty, a shop visitor offered him tickets to a series of lectures at the Royal Institution. (Although social mobility in Georgian London was very low, it wasn't quite zero.) Sir Humphry Davy was speaking on electricity, and on the hidden powers that must exist behind the surface of our visible universe. Faraday went, and realized he had been granted a lucky glimpse of a better life than he had working at the shop. But how could he enter it? He had not been to Oxford, or to Cambridge or, indeed, even attended much of what we call secondary school; he had as much money as his blacksmith father could give him—none—and his friends were just as poor.

But he could bind an impressive-looking book. Faraday had always been in the habit of taking notes when he could, and he'd brought back to the shop notes he'd taken at Davy's lectures. He wrote them out, and inserted a few drawings of Davy's demonstration apparatus. Then he rewrote the manuscript—all his drafts are kept today with the attention due a sacred relic, in the basement Archive Room of London's Royal Institution—took up his leather, awls, and engraving tools, and bound together a terrific book, which he sent to Sir Humphry Davy.

Davy replied that he wanted to meet Faraday. He liked him, and despite a disconcerting series of starts and stops, finally hired him away from the binder as a lab assistant.

Faraday's old shopmates might have been impressed, but his new position was not as ideal as he'd hoped. Sometimes Davy behaved as a warm mentor, but at other times, as Faraday wrote to his friends, Davy would seem angry, and push Faraday away. It was especially frustrating to Faraday, for he'd been drawn to science in large part by Davy's kind words; his hints that if only one had the skill and could see what's hitherto been hidden, everything we experience could actually be linked.

It took several years for Davy finally to ease up, and when he did it appeared to coincide with Faraday being asked to understand an extraordinary finding out of Denmark. Until then, everyone knew that electricity and magnetism were as unrelated as any two forces could be. Electricity was the crackling and hissing stuff that came from batteries. Magnetism was different; a totally invisible force that made navigators' needles tug forward, or drew pieces of iron to a lodestone. Magnetism was not anything you thought of as part of batteries and circuits. Yet a lecturer in Copenhagen had now found that if you switched on the current in an electric wire, any compass needle put on top of the wire would turn slightly to the side.

No one could explain this. How could the power of electricity in a metal wire possibly leap out and make a magnetic compass needle turn? Faraday was asked to work on how this link might occur.

Faraday, now in his late twenties, immediately cheered up. He started courting a girl ("You know me as well or better than I do myself," he wrote. "You know my former prejudices, and my present thoughts—you know my weaknesses, my vanity, my whole mind") and the girl liked being courted: in mid-1821, when Faraday was twenty-nine, they got married. He became an official member of the church which his family had been a part of for many years. This was a gentle, literalist group called the Sandemanians, after Robert Sandeman, who'd brought the sect to England. Most of all, Faraday now had a chance to impress Sir Humphry: to pay him back for his initial faith in hiring a relatively uneducated young binder, and to cross, finally, the inexplicable barriers Davy had raised between them.

Faraday's lack of limited formal education, curiously enough, turned out to be a great advantage. This doesn't occur often, because when a scientific subject reaches an advanced level, a lack of education usually makes it impossible for outsiders to get started. The doors are closed, the papers unreadable. But, in these early days of understanding energy, it was a different story. Most science students had been trained to show that any complicated motion could be broken down into a mix of straight lines, pushes and pulls. It was natural for them, accordingly, to try to see if there were any straight-line pulls between magnets and electricity. But this approach didn't show how the power of electricity might tunnel through space to affect magnetism.

Faraday, who did not have that bias of thinking in straight lines, turned to the Bible for inspiration. The Sandemanian religious group he belonged to believed in a different geometric pattern: the circle. Humans are holy, they said, and we all owe an obligation to one another based on our holy nature. I will help you, and you will help the next person, and that person will help another, and so on until the circle is complete. This circle wasn't merely an abstract concept. Faraday had spent much of his free time for years either at the church talking about this circular relation, or engaged in charity and mutual helping to carry it out.

He got to work studying the relationship between electricity and magnetism in the late summer of 1821. It was twenty years before Alexander Graham Bell, the inventor of the telephone, would be born; more than fifty years before Einstein. Faraday propped a magnet to stick up. From his religious background, he imagined a whirling tornado of invisible *circular* lines swirling around it. If he were right, then a loosely dangling wire could be tugged along, caught in those mystical circles like a small boat getting caught up in a whirlpool. He connected up the battery.

And immediately he had the discovery of the century.

Later, the apocryphal story goes—after all the announcements, after Faraday was made a Fellow of the Royal Society—the prime minister of the day asked what good this invention could be, and Faraday answered: "Why, Prime Minister, someday you can tax it."

What Faraday had invented, in his basement laboratory, was the basis of the electric engine. A single dangled wire, whirling around and around, doesn't seem like much. But Faraday had only a small magnet, and was feeding in very little power. Rev it up, and that whirling wire will still doggedly follow the circular patterns he had mapped out in seemingly empty air. Ultimately one could attach heavy objects to a similar wire, and they would be tugged along as well—that's how an electric engine works. It doesn't matter whether it is the featherweight spinning plate of a computer drive that's being dragged along, or the pumps that pour tons of fuel into a jet engine.

Faraday's brother-in-law, George Barnard, remembered Faraday at the moment of discovery: "All at once he exclaimed, `Do you see, do you see, do you see, George?' as the wire began to revolve. . . . I shall never forget the enthusiasm expressed in his face and the sparkling in his eyes!"

Faraday was sparkling because he was twenty-nine years old and had made a great discovery, and in a subtle way it really did seem to suggest that the deepest ideas of his religion were true. The crackling of electricity, and the silent force fields of a magnet— and now even the speeding motion of a fast twirling copper wire—were seen as linked together. As the amount of electricity went up, the available magnetism would go down. They were not separate events, but truly parts of a single unitary power. Faraday's invisible whirling lines were the tunnel—the conduit—through which magnetism could pour into electricity, and vice versa. The full concept of "Energy" had still not been formed, but Faraday's discovery that these different kinds of energy were linked was bringing it closer. It was the high point of Faraday's life—and then Sir Humphry Davy accused him of stealing the whole idea.

Davy began to let it be known that he had personally discussed the topic with a different researcher who had been investigating the topic—a properly educated researcher—and Faraday must have just overheard them.

The story was false, of course, and Faraday tried to protest, begging on the basis of their past friendship to let him explain, but Davy would have none of it. There were further crude hints, from others, if not from Davy himself: What else could you expect from a lower-class boy, from someone so junior, who was trying to wangle his way up as an apprentice; who knew nothing of what a more in-depth education could teach? After a few months Davy backed off, but he never apologized, and left the charges to dangle.

In notes and private journal entries Davy often wrote of how important it was to encourage young men. The problem was that he just couldn't bring himself to do it. The issue was nothing as simple as youth versus old age.

Davy was barely a decade older than Faraday. But Davy loved being lionized as the leader of British science, and all the time he spent away from the lab, soaking up praise in London high society with his status-keen wife, meant that the praise was increasingly false. He wasn't really on top of the latest research. When he corresponded with thinkers on the Continent he knew they were impressed at getting a letter from someone so prominent in the Royal Institution, but he avoided offering fresh ideas.

Hardly anyone else recognized this, but Faraday did. He was more like Davy than anyone else. Both men had started at a level much below that of their contemporaries in London science. Faraday made no excuses for that, but Davy did everything he could to hide his past. Faraday's quiet presence was a constant reminder of what they'd both once shared.

Faraday never spoke out against Davy. But for years after the charges of plagiarism and their repercussions, he stayed warily away from front-line research. Only when Davy died, in 1829, did he get back to work.

Faraday lived into old age, in time becoming prominent in the Royal Institution himself. His rise was typical of the move from gentlemanly to professional science. Davy's slurs against him were long forgotten. He went on to make other discoveries; he became very famous, and as often in demand, receiving such letters as this:

May 28th, 1850

Dear Sir,

It has occurred to me that it would be extremely beneficial to a large class of the public to have some account of your late lectures on the breakfast-table....I should be exceedingly glad to have ... them published in my new enterprise....

With great respect and esteem I am, Dear Sir, Your faithful servant,

Charles Dickens

$E=mc^2$ · $E=mc^2$ · $E=mc^2$

By the last decade of his life, though, Faraday—like Davy—was no longer able to follow the latest results. But the energy concept had taken on a life of its own. All the world's seemingly separate forces were slowly, majestically, being linked to create a masterpiece of the Victorian Age: the huge, unifying domain of Energy. Since Faraday had shown that even electricity and magnetism were linked—two items that had once seemed totally distinct—the scientific community was more confident that every other form of energy could similarly be shown to be deeply connected. There was chemical energy in an exploding gunpowder charge, and there was frictional heat energy in the scraping of your shoe, yet they were linked too. When a gunpowder charge went off, the amount of energy it produced in air blasts and falling rocks would be the same as what had been in the resting chemical charge inside.

It's easy to miss how extraordinary a vision was the energy concept that Faraday's work helped create. It's as if when God created the universe, He had said, I'm going to put X amount of energy in this universe of mine. I will let stars grow and explode, and planets move in their orbits, and I will have people create great cities, and there will be battles that destroy those cities, and then I'll let the survivors create new civilizations. There will be fires and horses and oxen pulling carts; there will be coal and steam engines and factories and even mighty locomotives. Yet throughout the whole thing, even though the types of energy that people see will change, even though sometimes the energy will appear as the heat of human or animal muscle, and sometimes it will appear as the gushing of waterfalls or the explosions of volcanoes: despite all those variations, the *total* amount of energy will remain the same. The amount I created at the beginning will not change. There will not be one millionth part less than what was there at the start.

Expressed like this it sounds like the sheerest mumbo-jumbo—Faraday's religious vision of a single universe, with just one single force spreading all throughout it. It's like something out of *Star Wars:* "The Force is the energy field created by all living things; it binds the galaxy together" (Obi-Wan Kenobi).

Yet it's true! When you swing closed a cupboard door, even if it's in the stillness of your home at night, energy will appear in the gliding movement of the door, but exactly that much energy was removed from your muscles. When the cupboard door finally closes, the energy of its movement won't disappear, but will simply be relocated to the shuddering bump of the door against the cupboard, and to the heat produced by the grinding friction of the hinge. If you had to dig your feet slightly against the floor to keep from slipping when closing the door, the earth will shift in its orbit and rebound upward by exactly the amount needed to balance that.

The balancing occurs everywhere. Measure the chemical energy in a big stack of unburned coal, then ignite it in a train's boiler and measure the energy of the roaring fire and the racing locomotive. Energy has clearly changed its forms; the systems look very different. But the total is exactly, precisely the same.

Faraday's work was part of the most successful program for further research the nineteenth century had seen. Every quantity in these energy transformations that Faraday and others had now unveiled could be computed and measured. When that was done, the results confirmed, always, that indeed the total sum had never changed—it was "conserved." This became known as the Law of the Conservation of Energy.

Everything was connected; everything neatly balanced. In the last decade of Faraday's life, Darwin seemed to have proven that God wasn't needed to create the living species on our planet. But Faraday's vision of an unchanging total Energy was often felt to be a satisfactory alternative: a proof that the hand of God really had touched our world, and was still active amidst us.

$E=mc^2$ · $E=mc^2$ · $E=mc^2$

This concept of energy conservation is what the science teachers in Einstein's cantonal high school in Aarau, in northern Switzerland, had taught him when he arrived there for remedial work in 1895, twenty-eight years after Faraday's death. Einstein had arrived at the school not because he'd had any desire to go there—he had already dropped out of one perfectly good high school in Germany, vowing that he'd had enough—but because he had failed his entrance exams at the Federal Institute of Technology in Zurich, the only university that offered a chance of taking a high school dropout. One kindly instructor there had thought he might have some merit, so instead of turning him away entirely, the institute's director had suggested this quiet school—set up on informal, student-centered lines—in the northern valleys.

When Einstein did finally make it into the Federal Institute of Technology—after his first delicious romance, with the eighteenth-year-old daughter of his Aarau host—the physics lecturers there were still teaching the Victorian gospel, of a great overarching energy force. But Einstein felt his teachers had missed the point. They were not treating it as a live topic, honestly hunting for what it might mean, trying to feel for those background religious intimations that had driven Faraday and the others forward. Instead, energy and its conservation was just a formalism to most of them, a set of rules. They didn't feel the need to go any deeper. There was a great complacency throughout much of

Western Europe at the time. European armies were the most powerful in the world; European ideas were "clearly" superior to those of all other civilizations. If Europe's top thinkers had concluded that energy conservation was true, then there was no reason to question them.

Einstein was easygoing about most things, but he couldn't bear complacency. He cut many of his college classes—teachers with that attitude weren't going to teach him anything. He was looking for something deeper, something broader. Faraday and the other Victorians had managed to widen the concept of energy until they felt it had encompassed every possible force.

But they were wrong.

Einstein didn't see it yet, but he was already on the path. Zurich had a lot of coffeehouses, and he spent afternoons in them, sipping the iced coffees, reading the newspapers, killing time with his friends. In quiet moments afterward, though, Einstein thought about physics and energy and other topics, and began getting hints of what might be wrong with the views he was being taught. All the types of energy that the Victorians had seen and shown to be interlinked—the chemicals and fires and electric sparks and blasting sticks—were just a tiny part of what might be. The energy domain was perceived as very large in the nineteenth century, but in only a few years Einstein would locate a source of energy that would dwarf what even the best, the most widely hunting of those Victorian scientists had found.

He would find a hiding place for further vast energy, where no one had thought to look. The old equations would no longer have to balance. The amount of energy God had set for our universe would no longer remain fixed. There could be more.

CHAPTER THREE

'='

Most of the main typographical symbols we use were in place by the end of the Middle Ages. Bibles of the fourteenth century had text that looked much like telegrams:

IN THE BEGINNING GOD CREATED THE HEAVEN AND THE
EARTH AND THE EARTH WAS WITHOUT FORM AND VOID
AND DARKNESS WAS UPON THE FACE OF THE DEEP

One change that took place over time was to drop most of the letters to lowercase:

In the beginning God created the heaven and the earth and the earth was without form and void and darkness was upon the face of the deep

Another shift was to insert tiny round circles to mark the major breathing pauses:

In the beginning God created the heaven and the earth. And the earth was without form and void and darkness was upon the face of the deep.

Smaller curves were used as well, for the minor breathing pauses:

In the beginning, God created the heaven and the earth ...

Major symbols were locked in rather quickly once printing began at the end of the 1400s. Texts began to be filled in with ? symbols and ! marks. It was a bit like Windows setting the standard in personal computers and driving out other operating systems.

Minor symbols took longer. By now we take them so much for granted that, for example, we almost always blink when we see the period at the end of a sentence. (Watch someone when they're reading and you'll see it.) Yet this is an entirely learned response.

For more than a thousand years, one of the world's major population centers used this symbol $\wedge$ for addition, since it showed someone walking toward you (and so was to be "added" to you), and $\wedge$ for subtraction. These Egyptian symbols could easily have spread to become universally accepted, just as other Middle Eastern symbols had done. The Hebrew א and ב—*aleph* and *beth*—were the source of the Greek α and β—*alpha* and *beta*—as in our word *alphabet*.

Through the mid-1500s there was still space for entrepreneurs to set their own mark by establishing the remaining minor symbols. In 1543, Robert Recorde, an eager textbook writer in England, tried to promote the new-style "+" sign, which had achieved some popularity on the Continent. The book he wrote didn't make his fortune, so in the next decade he tried again, this time with a symbol seemingly of his own creation that he was sure would take off. In the best style of advertising hype everywhere, he even tried to give it a unique selling point: ". . . And to avoide the tediouse repetition of these woordes: is equalle to: I will sette . . . a pair of parallels, or . . . lines of one lengthe, thus: ══════ bicause noe .2. thynges, can be moare equalle. . . ."

It doesn't seem that Recorde gained from his innovation, for it remained bitter competition with the equally plausible // and even with the bizarre [; symbol, which the powerful

German printing houses were trying to promote. The full range of possibilities proffered
at one place or another include, if we imagine
them put in the equation:

$$e \parallel mc^2$$
$$e \rightarrow mc^2$$
$$e .\text{æqus}. mc^2$$
$$e] [mc^2$$

There was even my favorite

$$e =\!\!=\!\!=\!\!=\!\!=\!\!= mc^2$$

Not until Shakespeare's time, a generation
later, was Recorde's victory finally certain.
Pedants and schoolmasters since then have often used the equals sign just to summarize
what's already known, but a few thinkers had a
better idea. If I say that *15+20=35*, this is not
very interesting. But imagine if I say:

(go **15** degrees west)
+
(then go **20** degrees south)
=
(you'll find trade winds that can fling you
across the Atlantic to a new continent in
35 days).

Then I am telling you something new. A good equation is not simply a formula for computation. Nor is it a balance scale confirming that two items you suspected were nearly equal really are the same. Instead, scientists started using the = symbol as something of a telescope for new ideas—a device for directing attention to fresh, unsuspected realms. Equations simply happen to be written in symbols instead of words.

This is how Einstein used the "=" in his 1905 equation as well. The Victorians had thought they'd found all possible sources of energy there were: chemical energy, heat energy, magnetic energy, and the rest. But by 1905 Einstein would say, No, there is another place you could look where you'll find more. His equation was like a telescope to lead there, but the hiding place wasn't far away in outer space. It was down here—it had been right in front of his professors all along.

He found this vast energy source in the one place where no one had thought of looking. It was hidden away in solid matter itself.

CHAPTER FOUR

'm' Is For Mass

For a long time the concept of "mass" had been like the concept of energy before Faraday and the other nineteenth-century scientists did their work. There were a lot of different material substances around—ice and tin and brick and rusted metal—but it was not clear how they related to each other, if they did at all.

What helped researchers believe that there had to be some grand links was that in the 1600s, Isaac Newton had shown that all the planets and moons and comets we see could be described as being cranked along inside an immense, God-created machine. The only problem was that this majestic vision seemed far away from the nitty-gritty of dusty, solid substances down here on earth.

To find out if Newton's vision really did apply on Earth—to find out, that is, if the separate types of substance around us really were interconnected in detail—it would take a person with a great sense of finicky precision; someone willing to spend time measuring even tiny shifts in weight or size. This person would also have to be romantic enough to be motivated by Newton's grand vision—for otherwise, why bother to hunt for these dimly suspected links between all matter?

This odd mix—an accountant with a soul that could soar—might have been a character portrait of Antoine-Laurent Lavoisier. He, as much as anyone else, was the man who first showed that all the seemingly diverse bits of tree and rock and iron on earth—all the "mass" there is—really were parts of a single connected whole.

$$E=mc^2 \cdot E=mc^2 \cdot E=mc^2$$

Lavoisier had demonstrated his romanticism in 1771 by rescuing the innocent thirteen-year-old daughter of his friend Jacques Paulze from a forced marriage to an uncouth, gloomy—yet immensely rich—ogre of a man. The reason he knew Paulze well enough to do the good deed for the daughter, Marie Anne, was that Paulze was his boss. The way he rescued Marie-Anne was to marry her himself.

It turned out to be a good marriage, despite the difference in age, and despite the fact that soon after the handsome twenty-eight-year-old Lavoisier rescued Marie Anne, he shifted back to being immersed in the stupendously boring accountancy work he did for Paulze, within the organization called the General Farm (la ferme generale).

It was not a real farm, but rather an organization with a near monopoly on collecting taxes for Louis XVI's government. Anything extra, the Farm's owners could keep for themselves. It was exceptionally lucrative, but also exceptionally corrupt, and for years had attracted old men wealthy enough to buy their way in, but unable to do any detailed accounting or administration. It was Lavoisier's job to keep this vast tax-churning device in operation.

He did that, head down, working long hours, six days a week on average for the next twenty years. Only in his spare time—an hour or two in the morning, and then one full day each week—did he focus on his science. But he called that single day his "jour de bonehur"—his "day of happiness."

Perhaps not everyone would comprehend why this was such a "bonheur." The experiments often resembled Lavoisier's ordinary accounting, only dragged out even longer. Yet the moment came when Antoine, in that irrational exuberance young lovers are known for, said his bride could now help him with a truly major experiment. He was going to watch a piece of metal slowly burn, or maybe just rust. He wanted to find out whether it would weigh more or less than it did before.

(Before going on, the reader might wish to actually guess: Let a piece of metal rust—think of an old fender or underbody panel on your car—and it ends up weighing

    a) less
    b) the same
    c) more

than it did before. Remember your answer.)

Most people, even today, probably would say it would weigh less. But Lavoisier, ever the cool accountant, took nothing on trust. He built an entirely closed apparatus, and he set it up in a special drawing room of his house. His young wife helped him: she was better at mechanical drawing than he was, and a lot better at English. (This would later be useful in keeping up with what the competition across the Channel was doing.)

They put various substances in their drawing room apparatus, sealed it tight, and applied heat or started an actual burn to speed up the rusting. Once everything had cooled down, they took out the mangled or rusty or otherwise burned-up metal and weighed it, and also carefully measured all the air rushing in or out.

Each time they got the same result. What they found, in modern terms, was that a rusted sample does not weigh less. It doesn't even weigh the same. It weighs *more*.

This was unexpected. The additional weight was not from dust or metal shards around in the weighing apparatus—he and his wife had been very careful. Rather, air has parts: there are different gases within the vapor we breathe. Some of the gases must have flown down and stuck to the metal. That was the extra weight he had found.

What was really happening? There was the same amount of stuff overall, yet now the oxygen that had been in the gases floating above was no longer in the air. *But it had not disappeared*. It had simply stuck on to the metal. Measure the air, and you would see it had lost some weight. Measure the chunk of metal, and you would see it had been enhanced—by exactly that same amount of weight the air had lost.

With his fussily meticulous weighing machine, Lavoisier had shown that matter can move around from one form to another, but yet will not burst in and out of existence. It was central to one of the prime discoveries of the 1700s—on a level with Faraday's realizations about energy in the basement of the Royal Institution a half century later. Here too, it was as if God had created a universe, and then said, I am going to put a fixed amount of mass in my domain, I will let stars grow and explode, I will let mountains form and collide and be weathered away by wind and ice; I will let metals rust and crumble. Yet throughout this the total amount of mass in my universe will never alter; not even to the millionth of an ounce; not even if you wait for all eternity. If a city were to be weighed, and then broken by siege, and its buildings burned by fire—if all the smoke and ash and broken ramparts and bricks were collected and weighed, there would be no change in the original weight. Nothing would have truly vanished, not even the weight of the smallest speck of dust.

To say that all physical objects have a property called their "mass," which affects how they move, is impressive, yet Newton had done it in the late 1600s. But to get enough detail to show exactly how their parts can combine or separate? That is the further step Lavoisier had now achieved.

Whenever France's scientists make discoveries at this level, they're brought close to the government. It happened with Lavoisier. Could this oxygen he'd helped clarify be used to produce a better blast furnace? Lavoisier was given a membership in the Academy of

Sciences, and funds to help find out. Could the hydrogen he was teasing out from the air with his careful measurements be useful in supplying a flotilla of balloons, capable of competing with Britain for supremacy in the air? He got grants and contracts for that as well.

In any other period this would have guaranteed the Lavoisiers an easy life. But all these grants and honors and awards were coming from the king, Louis XVI, and in a few years Louis was murdered, along with his wife and many of his ministers and wealthy supporters.

Lavoisier might have avoided being caught up with the other victims. The Revolution was only at its most lethal phases for a few months, and a lot of Louis's closest supporters simply lived out those periods in quiet. But Lavoisier could never drop the attitude of careful measuring. It was part of his personality as an accountant; it was the essence of his discoveries in science. Now it would kill him.

The first mistake seemed innocuous enough. Outsiders constantly bothered members of the Academy of Sciences, and long before the Revolution, one of them, a Swiss-born doctor, had insisted that only the renowned Lavoisier would be wise enough, and understanding enough, to judge his new invention. The device was something of an early infrared scope, allowing the doctor to detect the shimmering heat waves rising from the top of a candle, or of a cannonball, or even—on one proud occasion, when he'd lured the American representative to his chambers—from the top of Benjamin Franklin's bald head. But Lavoisier and the Academy turned him down. From what Lavoisier had heard, the heat patterns that the doctor was searching for couldn't be measured with precision, and to Lavoisier that was anathema. But the Swiss-born hopeful—Dr. Jean-Paul Marat—never forgot.

The next mistake was even more closely linked to Lavoisier's obsession with measurement. Louis XVI was helping America fund its revolutionary war against the British, an alliance that Benjamin Franklin had been central in sustaining. There were no bond markets, so to get the money Louis had to turn to the General Farm. But taxes already were high. Where could they go to get more?

In every period of incompetent administration France has suffered—and Louis's successors in the 1930s would have given him a good run—there almost always has been a small group of technocrats who've decided that since no one who was officially in power was going to take charge, then they would have to do it themselves. Lavoisier had an idea. Think of the measuring apparatus in his drawing room, the one where he and Marie Anne had been able to keep exact track of everything going in and out. Why not enlarge it, wider and wider, so that it encompassed all of Paris? If you could track the city's incomings and outgoings, he realized, you could tax them.

There once had been a physical wall around Paris, but it dated from medieval times, and had long since become nearly useless for taxation. Tollgates were crumbling, and many areas were so broken that smugglers could just walk in.

Lavoisier decided to build another wall, a massive one, where everyone could be stopped, searched, and forced to pay tax. It cost the equivalent, in today's currency, of several hundred million dollars; it was the Berlin Wall of its time. It was six feet high, of heavy masonry, with dozens of solid tollgates and patrol roads for armed guards.

Parisians hated it, and when the Revolution began, it was the first large structure they attacked, two days before the storming of the Bastille: they tore at it with firebrands and axes and bare hands till it was almost entirely gone. The culprit was known, as an antiroyalist broadsheet declared: "Everybody confirms that M. Lavoisier, of the Academy of Sciences, is the 'beneficent patriot' to

whom we owe the . . . invention of imprison-
ing the French capital. . . ."

Even this he might have survived. A mob's
passions are brief, and Lavoisier hurriedly tried
to show he was on their side. He personally
directed the gunpowder mills that supplied
the Revolutionary Armies; he tried to have the
Academy of Sciences show new, reformist cre-
dentials by getting rid of the grand tapestries
in its Louvre offices. He even seemed to be
succeeding—until one never-forgiving figure
from his past emerged.

By 1793, Jean-Paul Marat was head of a
leading faction in the National Assembly.
He'd suffered years of poverty because of
Lavoisier's rejection: his skin was withered
from an untreated disease, his chin unshaven,
his hair neglected. Lavoisier by contrast was
still handsome: his skin was smooth; his build
was strong.

Marat didn't kill him immediately. Instead,
he made sure Paris's citizens were constantly
reminded of the wall, this living, large-scale
summary of everything Marat hated about the
class-smug Academy. He was a magnificent
speaker—along with Danton and, in recent
history, Pierre Mendes-France, among the finest France has produced. ("I am the anger, the
just anger, of the people and that is why they
listen to me and believe in me.") The only sign

of Marat's tension—barely visible to listeners watching his confident posture, right hand on his hip, left arm casually extended on the desk in front of him—was a slight nervous tapping of one foot on the ground. As Marat denounced Lavoisier, he embodied the very principle that Lavoisier had demonstrated.

For was it not true that everything balances? If you seem to destroy something in one place, it's not really destroyed. It just appears somewhere else.

In November 1793 Lavoisier got word he was going to be arrested. He tried hiding in the abandoned parts of the Louvre, roaming through the Academy's empty offices there, but after four days he gave up, and walked— with Marie-Anne's father—to the Port-Libre prison.

If he looked out his window of the Port-Libre ("Our address is: first floor hall, number 23, room at the end"), he could see the great classical dome of the Observatory, a landmark over one century old, and now closed by Revolutionary orders. At least at night, when the guards ordered candles blown out in Lavoisier's prison, the stars were visible above its dome.

There were transfers to other prisons; the trial itself was on May 8. A few prisoners tried to speak, but the judges laughed at them.

Marat's bust was on a shelf looking down on the accused. That afternoon, twenty-eight of the onetime millionaires from the General Farm were taken to what's now the Place de la Concorde. Their hands were tied behind their backs. It was a steep climb up to the working level of Dr. Guillotin's instrument. Most seem to have been quiet, though one of the older men "was led to the scaffold in a pitiful state." Paulze was third; Lavoisier was fourth. There was about a minute after each beheading: not to clean the blades, but to clear away the headless bodies.

$$E=mc^2 \cdot E=mc^2 \cdot E=mc^2$$

With Lavoisier's work, the conservation of "mass" was on its way to being established. He had played a central role in helping to show that there was a vast, interconnected world of physical objects around us. The substances that fill our universe can be burned, squeezed, shredded, or hammered to bits, but they won't disappear. The different sorts floating around just combine or recombine. The total amount of mass, however, remains the same. It would be the perfect match to what Faraday later found: that energy is conserved as well. With all of Lavoisier's accurate weighing and chemical analysis, researchers were able to start tracing how that conservation happened in practice—as with his working out how oxygen molecules cascaded from the air to stick to iron. Breathing was more of the same, simply a means of shifting oxygen from the outer atmosphere to the inside of our bodies.

By the mid-1800s, scientists accepted the vision of energy and mass as being like two separate domed cities. One was composed of fire and crackling battery wires and flashes of light—this was the realm of energy. The other was composed of trees and rocks and people and planets—the realm of mass.

Each one was a wondrous, magically balanced world; each was guaranteed in some unfathomable way to keep its total quantity unchanged, even though the forms in which it appeared could vary tremendously. If you tried to get rid of something within one of the realms, then something else within that same realm would always pop up to take its place.

Nothing connected the two realms, however. There were no tunnels or gaps to get between the blocking domes. This is what Einstein was taught in the 1890s: that energy and mass were different topics; that they had nothing to do with each other.

Einstein later proved his teachers wrong, but not in the way one might expect. It is common to think of science as building up gradually from what came before. The tele-graph is tinkered with and turns into the tele-phone; a propeller airplane is developed, and studied, and then improved planes are built. But this incremental approach does not work with deep problems. Einstein did find a link between the two domains, but he didn't do it by looking at experiments with weighing mass and seeing if somehow a little bit was not accounted for, and might have slipped over to become energy. Instead he took what seems to be an immensely roundabout path. He seemed to abandon mass and energy entirely, and began to focus on what appeared to be an unrelated topic. He began to look at the speed of light.

CHAPTER
FIVE

"c" Is for celeritas

"c" is different from what we've looked at so far. "E" is the vast domain of energies, and "m" is the material stuff of the universe. But "c" is simply the speed of light.

The reason it has this unsuspected letter for its name is that it dates from the period xxx when science was centered in Italy, and Latin was the language of choice. *Celeritas* is the Latin word meaning "swiftness" (and the root of our word *celerity.*

This chapter will look at how "c" came to play such an important role in $E=mc^2$: how this particular speed—what might seem an arbitrary number—can actually control the link between all the mass and all the energy in the universe.

For a long time even measuring the speed of light was considered impossible. Almost everyone was convinced that light traveled infinitely fast. But if that were so, it could never have been used in a practical equation. Before anything more could be done—before Einstein could have possibly thought of using "c"—someone had to confirm that light travels at a finite speed, but that wouldn't be easy.

$E=mc^2$ · $E=mc^2$ · $E=mc^2$

Galileo was the first person to try measuring the speed of light, while he was undergoing house arrest in his old age, nearly blind. He was too old to carry out the experiment himself, and the Inquisition had given strict orders that he was to never go outdoors. But that was little more than a challenge to him and his friends. When members of the Florentine academy heard of his work, they let it be known that they would do the observations for him.

The idea was as simple as all Galileo's work had been. Two volunteers were to stand holding lanterns on hillsides a mile apart one summer evening. They would open their lantern shutters one after the other, and then time how much of a delay there was for the light to cross the valley.

The experiment was a good idea, but the technology of the time was too poor to get any clear result. Galileo had trained himself to

breathe regularly, so as not to speed his heart-
beat when an experiment was under way, for
he used his pulse to measure short intervals of
time. But that summer evening, in the hills
outside of Florence, the volunteers found the
light was too quick. All they noticed was a
quick blur, a movement that seemed instantaneous. This could have been seen as a failure,
and for most people it was just another proof
that light traveled at infinite speed. But when
Galileo heard of the Florentines' failure, he
didn't accept that it meant his speculations
were wrong. Rather, he concluded, it would
just have to be left to someone from a future
generation to find a way to time this impossibly fast burst.

In xxx, several decades after Galileo's death
in 1642, Jean-Dominique Cassini arrived in
Paris to take up his position as head of the
newly established Paris Observatory. There
was a lot of new construction to supervise, and
he could sometimes be seen in the street doing that—not far from the shadows of the
Port-Libre prison, where Lavoisier in the next
century would await his death—but his most
important task was to shake some life into
French science. He also had a personal incen-
tive to make the new institution succeed, for
his name wasn't actually Jean-Dominique but
Giovanni Domenico. And he wasn't French,

but newly arrived from Italy, and although the king was on his side, and the funding was said to be guaranteed, who knew how long that really would last?

Cassini sent emissaries to recruit staff from other observatories, and especially from the fabled observatory (It was also an occasion to accurately fix the coordinates of Uraniborg, to help in measuring distances for navigation.) The founder of the Uraniborg observatory, Tycho Brahe, had made the observations on which Kepler and even Newton based their work. Brahe had created unimagined luxuries: there were exotic species of trees, gardens with dozens of fish ponds around the central castle; an impressive intercomlike communication system; there were even rumors of an automatic flush toilet.

Cassini's right-hand man, Jean Picard, reached Uraniborg in 1671, sailing the misty waters from Copenhagen. He was excited about finally getting to see the fabled stronghold—then dismayed at finding it was a complete wreck.

Those sophisticated findings that had impressed Kepler dated from almost a century before. The observatory's founder had been a powerful personality, but when he had died, no one really took over. Everything was decayed or broken when Picard arrived: the fish ponds filled in, the quadrants and celestial globe rusted; the intercom tubes suitable just for mice.

The only thing Picard did manage to bring back to Paris was a bright twenty-year-old Dane named Ole Roemer. Others might be humbled to meet the great Cassini when they arrived back, for Cassini was the world authority on the orbits of the planet Jupiter, and of its satellites as they rotated around the planet. But although today we think of Denmark as a small nation, at that time it ran an empire that encompassed a good stretch of northern Europe. Roemer was proud, and confident, and set out to make his own name.

It's doubtful whether Cassini was especially pleased with the upstart. It had taken a long time for him to make the switch from Giovanni Domenico to Jean-Dominique. He had accumulated numerous detailed observations of the planets, and he was certainly going to use them to maintain his worldwide reputation. But what if Roemer plundered his findings to prove that the conclusions Cassini was drawing from them were all wrong?

The reason this was possible was that there was a problem with the innermost moon of Jupiter, the one called Io. It was supposed to travel around its planet every 42½ hours. But it never stuck honestly to schedule. Sometimes it was a little quicker, sometimes a little

slower. There was no discernible pattern anyone could.

But why? To solve the problem, Cassini insisted on more measurements. The effort
might be exhausting for the observatory's director, and of course it would entail more staff
and more equipment and more funds and
more patronage, and all that embarrassing
public exposure, but if that was necessary, it
could be done. To Roemer, however, what
was needed was not the sort of complex measurements only middle-aged administrators
could manage. What was needed was the bril-
liance, the inspiration, which a young outsider
applying his mind could provide.

And this is what Roemer did. Everyone—
even Cassini—assumed that the problem was
in how Io traveled. Possibly it was ungainly
and wobbled during its orbit; or possibly there
were clouds or other factors around Jupiter
that slowed Io unevenly. Roemer reversed the
problem. Cassini had made observations of Io,
and the observations showed that something
about its orbit was not smooth. But why
should the flaw be assumed to rest far away
near Jupiter? The data to look at was not how
Io was traveling, thought Roemer. The question was how Earth was traveling.

To Cassini, this couldn't possibly matter.
Like almost everyone else, he was convinced
that light traveled as an instantaneous flash.

Any fool could see that. Hadn't Galileo's own experiment shown that there was no evidence to the contrary?

Roemer ignored all that. Suppose—just suppose—that light did take some time to travel the great distance from Jupiter. What would that mean? Roemer imagined he was straddling the solar system, watching the first flicker of Io's light burst out from behind the planet Jupiter, and rush all the way to Earth. In the summer, for example, if Earth was closer to Jupiter, the light's journey would be shorter, and Io's image would arrive sooner. In the winter of the same year, though, Earth could have swung around to the other side of the solar system. It would take a lot longer for Io's signal to reach us.

Roemer went through Cassini's stacked years of observations, and by the late summer of 1676 he had his solution: not just a hunch, but an exact figure for how many extra minutes light took to fly that extra distance when Earth was far from Jupiter.

What should he do with such a finding? By protocol, Roemer should have let Cassini present it as his own work, and simply nod modestly, perhaps, when the observatory chief paused to remark that he couldn't have done it without the help of this young man whose future career was worth watching.

Roemer didn't go along with that. In September, at a public meeting of the new Academy of Sciences, he proclaimed a challenge. Astronomy is an exact science, and even seventeenth-century tools were good enough to determine that the satellite Io was scheduled to appear from behind Jupiter on the coming November 9, sometime in the late afternoon. By Cassini's reasoning, it would be detected at 5:25 P.M. on that day. That was what he extrapolated from when it had last been clearly sighted, in August.

Roemer declared that Cassini was going to be wrong. In August, he explained, the earth had been close to Jupiter. In November it would be further away. There would be nothing visible at 5:25—the light, though fast, would still be on its way, since it had to travel that extra distance. Even at 5:30 it wouldn't have made it across the solar system. Only at 5:35—no, 5:35 and 45 seconds, precisely— would anyone be able to get their first sighting on November 9.

There are many ways to make astronomers happy. A new supernova is good; a renewed grant from the government is good; tenure is extremely good. But an out-and-out battle between two of your distinguished colleagues? It was heaven. Roemer had thrown down his challenge partly out of pride, but also because he knew that Cassini was a much better political operator than he was. Roemer would only be able to claim credit if his prediction was so clear that Cassini and his minions couldn't wangle out of it if they were wrong.

The prediction was read out in September. On November, observatories from Greenwich to Milan had their telescopes ready. 5:25 P.M. arrived. No Io.

5:30 arrived. Still no Io.

5:35 P.M.

And then it appeared, at 5:35 and 45 seconds exactly.

And Cassini declared he had not been proven wrong! (Spin-doctoring was not invented in the era of television.) Cassini had lots of supporters, and support him they did. Who'd ever said Io was expected at 5:25? That had only been Roemer, they now declared. Besides, everyone realized Io's arrival time was never certain. It was so far away, so hard to see exactly, that perhaps it was traveling unevenly, or maybe clouds from Jupiter's upper atmosphere made a distorting haze. Who knew?

In the usual history of science accounts, it's not supposed to happen this way. Roemer had performed an impeccable experiment, with a clear prediction, yet Europe's astronomers still did not accept that light traveled at a finite speed. Cassini's supporters had won: the official line remained that the speed of light was just a mystical, unmeasurable figure; that it should have no impact on astronomical measurements.

Roemer gave up, and went back to Denmark, spending many years as the director of the port of Copenhagen. Only fifty years later—after a further generation had passed, and Jean-Dominique Cassini was gone—did further experiments convince astronomers that Roemer had been right. The value he had estimated for light's speed was close to the best current estimate, which is about 670,000,000 mph. (In fact the exact speed is 670,214,995 miles per hour, but for convenience we'll round it off to 670 million for the rest of this book.)

To emphasize how big a number this is, at 670,000,000 mph, you could get from London to Los Angeles in under 1/20th of a second. That explains why Galileo's experiment could not detect the time it took light to cross a valley outside Florence; the distance was much too small.

Here's another comparison. Mach 1 is the speed of sound, about 700 mph. A 747 jet travels at a little under Mach 1. The space shuttle, after full thrust, can reach Mach 25.

The asteroid or comet that splashed a hole in the ocean floor and destroyed the dinosaurs impacted at about Mach 70.

The number for "c" is Mach 900,000.

This vast speed leads to many curious effects. For example, let someone irritatingly speak into a cell phone just a few tables away from you in a restaurant; and it seems as if you're hearing his voice almost as soon as the words leave his mouth. But air only can carry a sound wave at the lowly speed of Mach 1, whereas radio signals shooting upward from the cell phone travel as fast as light. The person on the receiving end of the phone—even if she's hundreds of miles away—will hear the words *before* they've trundled the few yards through the air in the restaurant to reach you!

$E=mc^2 \cdot E=mc^2 \cdot E=mc^2$

To see why Einstein chose *this* particular value for his equation, we need to look closer, at light's inner properties. The story leaves the epoch of Cassini and Roemer far behind, and picks up in the late 1850s, the period just before the American Civil War, when the now elderly Michael Faraday began to correspond with James Clerk Maxwell, a slender young Scot still in his twenties.

It was a difficult time for Faraday. He could often barely get through a morning without extensive notes to remind him of what he was supposed to do. Even worse, Faraday also knew that the world's great physicists, almost all of whom had gone to elite universities, still patronized him. They accepted his practical lab findings, but few of them took seriously his theoretical interpretations, which he felt—but, with his lack of mathematics, could never prove—were the only way to explain what he'd found. To standard physicists, when electricity flowed through a wire it was basically like water flowing through a pipe: once the underlying mathematics was finally worked out, they believed, it would not be too different from what Newton and his numerous mathematically astute successors could describe.

Faraday, however, still went on about those strange circles and other wending lines from his religious upbringing. The area around an electromagnetic event, Faraday held, was filled with a mysterious "field," and stresses within that field produced what are interpreted as electric currents and the like. He insisted that sometimes you could almost see their essence, as in the curving patterns that iron filings take when they are sprinkled around a magnet. Yet almost no one believed him—except, now, for this young Scot named Maxwell.

At first glance the two men seemed very different. In his many years of research, Faraday had accumulated over 3,000 paragraphs of dated notebook entries on his experiments, from investigations that began early every morning. Maxwell, however, quite lacked any ability to get a timely start to the day. (When he was told that there was mandatory 6 A.M. chapel at Cambridge University, the story goes that he took a deep breath, and said, "Aye, I suppose I can stay up that late.") Maxwell also had probably the finest mathematical mind of any nineteenth-century theoretical physicist, while Faraday had problems with any conventional math much beyond simple addition or subtraction.

But on a deeper level the contact was close. Although Maxwell had grown up in a great baronial estate in rural Scotland, the family name had until recently simply been Clerk, and it was only from an inheritance on the maternal side that they'd acquired the more distinguished Maxwell to tack on. When young James was sent away to boarding school in Edinburgh, the other children— stronger in build, cockily confident with their big-city ways—picked on him: week after week, year after year. James never expressed any anger about it; just once, he quietly remarked: "They never understand me, but I understood them." Faraday also still carried the wounds from his experiences with Sir Humphry Davy in the 1820s, and would relapse into a quiet, watching solitude almost instantly after he'd finished an evening as an apparently ebullient speaker at one of the Royal Institution public lectures.

When the young Scot and the elderly Londoner corresponded, and then later when they met, they cautiously made contact of a sort they could share with almost no one else. For beyond the personality similarities, Maxwell was such a great mathematician that he was able to see beyond the surface simplicity of Faraday's sketches. It was not the childishness that less gifted researchers mocked it for. ("As I proceeded with the study of Faraday, I perceived that his method . . . was also a mathematical one, though not exhibited in the conventional form of mathematical symbols.") Maxwell took those crude drawings of invisible force lines seriously. They were both deeply religious men; they both took this possibility of God's immanence in the world very earnestly.

Back in his 1832 breakthrough, and then in much research after, Faraday had shown ways in which electricity can be turned into magnetism, and vice versa. In the late 1850s, Maxwell extended that idea, into the first full explanation of what Galileo and Roemer had never understood.

What was happening inside a light beam, Maxwell began to see, was just another variation of this back-and-forth movement. When a light beam starts going forward, one can think of a little bit of electricity being produced, and then as the electricity moves forward it powers up a little bit of magnetism, and as that magnetism moves on, it powers up yet another surge of electricity, and so on like a braided whip snapping forward. The electricity and magnetism keep on leapfrogging over each other in tiny, fast jumps—a "mutual embrace," in Maxwell's words. The light Roemer had seen hurtling across the solar system, and which Maxwell saw bouncing off the stone towers at Cambridge, was merely a sequence of these quick, leapfrogging jumps.

It was one of the high points of nineteenth-century science; Maxwell's Equations summarizing this insight became known as the greatest theoretical achievement between the work of Newton in the seventeenth century and that of Einstein in the twentieth century. Faraday never grasped the detailed mathematics behind Maxwell's expansions of his old work, and once even wrote to Maxwell, yearning for a simpler explanation. Maxwell was always kind to Faraday, but he too was slightly dissatisfied with what he'd produced. For how exactly did this strangely leapfrogging light wave braid itself along? No one could say for sure.

$E=mc^2$ · $E=mc^2$ · $E=mc^2$

Einstein's genius was to look closer at what this meant, even though he had to do it largely on his own. He had the confidence to do this: his final high school preparation in Aarau really had been superb, and he'd grown up to believe he could always question authority. By the 1890s, when Einstein was a student, Maxwell's formulations were usually taught as a received truth. But Einstein's main professor at the Zurich polytechnic, unimpressed with theory in physics, refused even to teach Maxwell to his undergraduates. (It was Einstein's resentment at being treated like this that led him to address that professor mockingly as *Herr* Weber rather than the expected *Herr Professor Weber*—a slight that Weber avenged by refusing to write a proper letter of recommendation for Einstein, leading to his years of isolation at the patent office job.)

When Einstein cut classes to go to the coffeehouses in Zurich, it was often with his own copies of Maxwell's work in hand. But, as for Maxwell, something about the leapfrogging bothered Einstein. If light was a wave like any other, he asked, then if you ran after it, could you catch up?

An example from surfing will show the problem. When you're first out in the water, trying not to let everyone on shore see how scared you are, the waves slosh past you. But once you force yourself to stand up on your surfboard, you can glide shoreward as the wave of water pushing you seems to be standing still around you. Be bold enough—or foolhardy enough—to do this in the extreme surf off Hawaii, and an entire curving tube of water might seem to be at rest around above and beside you.

Only in 1905 did the full insight hit Einstein. Light waves are different from everything else. A surfer's water wave can appear to hold still, because all the parts of the wave take up a steady position in relation to one another. That's why you can glance out from your surf board, and see a hovering sheet of water. Light is not like that, however. Light waves keep themselves going only by virtue of one part moving forward and so powering up the next part. (The electricity part of the light wave shimmers forward, and that "squeezes" out a magnetic part; then that magnetic part, as it powers up, creates a further "chunk" of electricity so the rushing cycle starts repeating.)

Whenever you think you're racing forward fast enough to have pulled up next to a light beam, look harder and you'll see that whatever part you thought you were close to is powering up a further part that is still hurtling away from you.

To catch up with a streak of light and see it standing still would be like saying, "I want to see the blurred arcs of a thrilling juggling act, but only if the balls are not moving." You can't do it. The *only* way you'll see a blur from the juggling balls is if they're moving fast.

Einstein concluded that light can exist only when a light wave is actively moving forward. It was an insight that had been lurking in Maxwell's work for over forty years, but no one had recognized it.

This new realization about light changed everything. The speed of light becomes the fundamental speed limit in our universe: nothing can go faster.

What we see as light only arises from the "blur" of the interchanging electricity and magnetism pumping each other forward. That's why light can't exist when it's standing still, and why we can never catch up with it.

It's easy to misunderstand this. If you were traveling at 669,999,999 mph, couldn't you pump in more fuel, and go the few mph faster—to 670,000,000, and then to

670,000,0001—to take you past speed of light? But the answer is that you can't, and it's not a quirk about the present state of earthly technology.

A good way to recognize this is to remember that light isn't just a number, but is a physical process. There's a bit difference. If I say that -273 (negative 273) is the lowest number that there is, you could rightly answer that I'm wrong: that -274 is lower, and -275 is lower yet, and that you can keep on going forever. But suppose we were dealing with temperatures. The temperature of a substance is a readout of how much its inner parts are moving, and at some point they're going to stop vibrating entirely. That happens at about -273 degrees on the Centigrade scale, and that's why -273 degrees is said to be "absolute zero" when you're talking about temperature. Pure numbers might be able to go lower, but physical things can't: a coin or a snowmobile or a mountain can't vibrate any less than not vibrating at all.

So it is with light. The 670,000,000 mph figure that Roemer measure for the light speeding down from Jupiter is a statement about what that light is like. It's a physical "thing." Light will always be a quick leapfrogging of electricity out from magnetism, and then of magnetism leaping out from electricity, all swiftly shooting away from anything trying to catch up to it. That's why its speed can be an upper limit.

$$E=mc^2 \cdot E=mc^2 \cdot E=mc^2$$

It's an interesting enough observation, but a cynic might say, so what if there *is* an ultimate speed limit? How should that affect all the solid objects that move around within the universe? You can clamp a label saying "Warning: No Speed Over 670,000,000 mph Can Be Achieved," on the signs by a busy road, but the traffic whirring past will be unaffected.

Or will it? This is where Einstein's whole argument finally circles back: where he showed that light's curious properties—the fact that it inherently squiggles away from you, and is therefore the ultimate speed limit—really enters into the nature of energy and mass. An example updated from one that Einstein himself used can suggest how it might happen.

Suppose a super space shuttle is blasting along very close to the speed of light. Under normal circumstances, when that space shuttle is going slowly, the fuel energy that's pumped into the engines would just raise its speed. But things are different when the shuttle is right at the very edge of the speed of light. It can't go much faster.

The pilot of the shuttle doesn't want to accept this, and starts frantically leaping up and down on the thruster control to get the vessel to go faster. But of course the pilot sees any light beam that's ahead still squirting out of reach at the full speed of "c." That's what light does for any observer. Despite the pilot's best efforts, the shuttle is not gaining on it. So what happens?

Think of frat boys jammed into a phone booth, their faces squashed hard against the glass walls. Think of a parade balloon, with a blowing air pump cabled into it that can't be turned off. The whole balloon starts swelling, far beyond any size for which it was intended. The same thing would happen to the shuttle. The engines are roaring with energy but that can't raise the shuttle's speed, for nothing goes faster than light. but the energy can't just disappear, either.

As a result, the energy being pumped through gets "squeezed" into becoming mass. Viewed from outside, the solid mass of the shuttle starts to grow. There's only a bit of swelling at first, but as you keep on pouring in energy, the mass will keep on increasing. The shuttle will keep on swelling.

It sounds preposterous, but there's evidence to prove it. If energy is poured into small protons, which have one "unit" of mass when they're standing still, they'll start moving faster and faster, till they're speeding very close to the speed of light. An observer really will see the protons begin to change—it's a regular event at the accelerators outside of Chicago, and at CERN (the European center for nuclear research) near Geneva, and everywhere else physicists do this work. The protons first swell to become two units of mass—twice as much as they were at the start—then three units, then on and on, as the power continues to be pumped in. At speeds of 99,9997 percent of "c," the protons end up 430 times bigger than their original size. So much power is drained from nearby electricity stations that the main experiments are often scheduled to run late at night, so that nearby residents won't complain about their lights dimming.

Energy that's pumped into the protons or into our imagined shuttle will turn into extra mass. Just as the equation states: that "E" can become "m," and "m" can become "E."

That explains why the "c" is in the equation. In this example, as you push up against the speed of light, that's where the linkages between energy and mass become especially clear. The figure "c" is merely a conversion factor telling you how that linkage operates.

Whenever you link two systems that developed separately, you need a conversion factor. To go from Centigrade to Fahrenheit, you multiply the Centigrade figure by 9/5, and then add 32. To go from centimeters to inches there's another rule: you multiply the centimeters by 0.3937.

The conversion factors seem arbitrary, but that's because they link measurement systems that evolved separately. Inches, for example, began in medieval England, and were based on the length of the human thumb tip. Thumbs are excellent portable measuring tools, since even the poorest individuals could count on regularly carrying them along to market. Centimeters, however, were popularized centuries later, during the French Revolution, and are defined as one billionth of the distance from the Equator to the North Pole, passing by Paris. It's no wonder the two systems don't fit together smoothly.

For centuries, energy and mass had also seemed to be entirely separate things. They evolved without contact. Energy was thought of as horsepower or kilowatt hours; mass was measured in pounds or kilos or tons. No one thought of connecting the units. No one glimpsed what Einstein did, that there would be a "natural" transfer between energy and mass, as we saw with the shuttle example, and that "c" is the conversion factor linking the two.

The reader might wonder when we'll get to the theory of relativity. The answer is that we already have been using it! All these points about a speeding shuttle and its expanding mass are central to what Einstein published in 1905.

Einstein's work changed the two separate visions scientists had taken from the nineteenth-century work on conservation laws. Energy isn't conserved, and mass isn't conserved—but that doesn't mean there is chaos. Instead, there's actually a deeper unity, for "something" is controlling the link between what happens in the energy domain and what happens in the seemingly distinct mass domain. The amount of mass that's gained is always going to be balanced by an equivalent amount of energy that's lost.

Lavoisier and Faraday had seen only part of the truth. Energy does not stand alone, and neither does mass. But the sum of mass plus energy will always remain constant.

This, finally, is the ultimate extension of the separate conservation laws the eighteenth- and nineteenth-century scientists had once thought complete. The reason this effect had remained hidden, unsuspected, all the time before Einstein, is that the speed of light is so much higher than the ordinary motions we're used to. The effect is weak at walking speed, or even at the speed of locomotives or jets, but it's still there. And as we'll see, the linkage is omnipresent in our ordinary world: all the energy is held quiveringly ready inside even the most ordinary substances.

Linking energy and mass via the speed of light was a tremendous insight, but there's still one more detail to get clear. A famous Sydney Harris cartoon shows Einstein at a board, trying out one possibility after another: $E=mc^1$, $E=mc^2$, $E=mc^3$, . . . But he didn't really o it that way, arriving at the squaring of "c" by mere chance.

So why did the conversion factor turn out to be c^2?

CHAPTER SIX

2

Enlarging a number by "squaring" it is an ancient procedure. A garden that has four paving slabs on one edge, and four on the other, doesn't have eight stones in it. It has 16.

The convenient shorthand that summarizes this action of building up a "square"—of multiplying a number by itself—went through almost the same range of permutations as did the Western typography for the equals sign.

But why should it appear in physics equations? The story of how an equation with a "squared" in it came to be plucked from all other possibilities for representing the energy of a moving object takes us back to France once more—to the early 1700s—and the generation halfway between Roemer and Lavoisier.

In February 1726, the thirty-one-year-old playwright Francois-Marie Arouet was convinced he'd successfully gate-crashed the establishment in France. He'd risen from the provinces to receive grants from the king, acceptance at the homes of noblemen, and in fact one evening was being dined at the gated home of the Duc de Sully. A servant interrupted the meal: there was a gentleman outside to see Arouet.

He went out, and probably had a moment to recognize the carriage of Duc de Rohan, an unpleasant, yet staggeringly rich man whom he'd teased in public when they were attending a play at the Comedie Francaise recently.

Then de Rohan's bodyguards got to work, beating Arouet while de Rohan watched, delighted, from inside his carriage, "supervising the workers," as he later described it. Somehow, Arouet managed to get back inside the gates, and into de Sully's home. But instead of sympathy or even outrage, Arouet encountered only laughter. De Sully and his friends were amused: a preposterous wordsmith had been put in his place by someone who really mattered.

Arouet vowed to avenge himself; he would challenge de Rohan to a duel, and kill him.

That was getting too serious. De Rohan's family had a word with the authorities to avert it; there was a police hunt; Arouet was arrested, then put in the Bastille.

When he finally got out he crossed the Channel, falling in love with England, and especially—estate agents take note—with the bucolic wonderland of Wandsworth, far from the grime of the busy city. He was exhilarated to find that there was a new concept in the air, the works of Newton, which represented what could be the opposite of the ancient, locked-in aristocratic system he'd known in France.

Newton had created a system of laws that seemed to detail, with superb accuracy, how every part of our universe moved about. The planets swung through space at a rate and in directions that Newton's laws described; a cannonball fired in the air would land exactly where Newton's calculations of its trajectory showed that it would land.

It really was as if we were living inside a vast windup clock, and all the laws Newton had seen were simply the gears and cogs that made it work. But if we could demand a rational explanation of the grand universe beyond our planet, Arouet wondered, shouldn't we be able to demand the same down here on earth? France had a king who demanded obedience, on the grounds that he was God's regent on earth. Aristocrats got authority from the king, and it was impious to question this. What if the same kind of analysis used in science by Newton could be used to reveal the role of money or vanity or other hidden forces in the political world as well?

By the time Arouet went back to Paris, three years later, he had begun pushing his new ideas, in private letters and printed essays. In a world of clear, levelheaded analysis of true forces, his humiliation outside the gates of de Sully would never had been allowed. Arouet would support Newton's new vision accordingly his whole life long. He was a good supporter to have, for Arouet was only the name he'd been born with. He'd already largely put it to the side for the pen name by which he was better known: Voltaire.

But even a skilled writer, however eager to push a particular thinker, can't shift a nation on his own. Voltaire needed to be able to place his talents within a switching center that could multiply them. The king's Academy of Sciences was too backward-looking; too locked into the old guard's way of thinking. The salons of Paris wouldn't do either. The usual hostesses were rich enough to keep a tame poet or two ("If you neglect to enroll yourself among the courtesans," Voltaire observed, "you are . . . crushed"), but there was no space for a real thinker. He needed help. And he found it.

He'd actually met her without realizing it, fifteen years before, visiting her father when she was just a girl. Emilie de Breteuil's family lived overlooking the Tuileries gardens in Paris, in an apartment with thirty rooms and seventeen servants. But although her brothers and sisters turned out as expected, Emilie was different, as her father wrote: "My youngest flaunts her mind, and frightens away the suitors. . . . We don't know what to do with her."

When she was sixteen they brought her to Versailles, but still she stood out. Imagine the actress Geena Davis, Mensa member and one-time action-film star, trapped in the early eighteenth century. Emilie had long black hair and a look of perpetual startled innocence, and although most other debutante types wanted nothing more than to use their looks to get a husband, Emilie read Descartes's analytic geometry, and wanted potential suitors to keep their distance.

She'd been a tomboy as a child, loving to climb trees, and she was also taller than average, and—best of all—since her parents had been worried she'd end up clumsy, they'd paid for fencing lessons for years. She challenged Jacques de Brun, whose position was roughly equivalent to head of the king's bodyguard detail, to a demonstration duel, in public, on the fine parquet of the great Hall of Arms. She was fast enough, and strong enough, with the thrusts and parries, to remind any overeager suitors that they would be wise to leave her alone.

Her intellect left her isolated at Versailles, for there was no one she could share her excitement with about the wondrous insights she was discovering through the work of Descartes and other researchers. (At least there were certain advantages in being immersed in equations—she found it easy to memorize cards at the blackjack table.)

When Emilie was nineteen, she chose one of the least objectionable courtiers as a husband. He was a wealthy soldier named du Châtelet, who would conveniently be on distant campaigns much of the time. It was a pro forma arrangement, and in the habit of the time, her husband accepted her having affairs while he was away. There were a number of lovers, ending with a onetime guards officer, Pierre-Louis Maupertuis, who had resigned his post, and was in the process of becoming a top physicist. Their courtship had begun in studying calculus and more advanced work together, but he was leaving on a polar expedition, and in 1730s France, no twenty-something young woman—however bright, however athletic—would be allowed to go along.

Now Emilie was at loose ends. Where could she turn for warmth? She had a few desultory affairs while Maupertuis was ordering his final supplies, but who, in France, could fill Maupertuis's place? Enter Voltaire.

"I was tired of the lazy, quarrelsome life led at Paris," Voltaire recounted later, ". . . of the privilege of the king, of the parties and cabals among the learned. . . . In the year 1733 I met a young lady who happened to think nearly as I did. . . ."

She met Voltaire at the opera, and although there was some overlap with Maupertuis, that was no problem. Voltaire composed a stirring poem for Maupertuis, complimenting him as a modern-day argonaut, for his boldness in venturing to the far north for science; he then wrote a romantic poem to du Châtelet, comparing her to a star, and noting that he, at least, was not so faithless as to exchange her for some expedition to the Arctic pole. It wasn't entirely fair to Maupertuis, but du Châtelet didn't mind. Anyway, what could Voltaire do? He was in love.

And so, finally, was she. This time she wasn't going to let it go. She and Voltaire shared deep interests: in political reform, in the fun of fast conversation ("she speaks with great rapidity," one of her earlier lovers had written, ". . . her words are like an angel"); above all, they shared a drive to advance science as much as they could. Her husband had a chateau, at Cirey, in northeastern France. It had been in the family since before Columbus went to America, and now was largely shuttered up; abandoned. Why not use that as a base for genuine scientific research in France? They got to work, and Voltaire soon wrote to a friend that Mme du Châtelet

> . . . is changing staircases into chimneys and chimneys into staircases. Where I ordered the workmen to construct a library, she tells them to place a salon. . . . She's planting lime trees where I'd planned to place elms, and where I only planted herbs and vegetables . . . only a flowerbed will make her happy.

Within two years it was complete. There was a library comparable to that of the Academy of Sciences in Paris, the latest laboratory equipment from London, and there were guest wings, and the equivalent of seminar areas, and soon there were visits from the top researchers in Europe. Du Châtelet had her own professional lab, but the wall decorations in her reading areas were original paintings by

Watteau; there was a private wing for Voltaire, yet also a discreet passageway conveniently connecting his bedroom with hers. (Using it one time when she didn't expect it, he discovered her with another lover, and she tried putting him at ease by explaining that she'd only done this because she knew he hadn't been feeling well, and hadn't wanted to trouble him while he needed rest.)

The occasional visitors from Versailles who came to scoff saw a beautiful woman willingly staying inside, working at her desk well into the evening, twenty candles around her stacks of calculations and translations; advanced scientific equipment stacked in the great hall. Voltaire would come in, not merely wanting to gossip about the court—though, being Voltaire, he was unable to resist this entirely—but also to compare Newton's Latin texts with some of the latest Dutch commentaries.

At several times she came close to jump-starting future discoveries. She performed a version of Lavoisier's rust experiment, and if the scales she'd been able to get machined had been only a bit more accurate, she might have been the one to come up with the law of the conservation of mass, years before Lavoisier was born.

The Cirey team kept up a supporting correspondence with other new-style researchers;

supplying them with whatever evidence, diagrams, calculations might be needed. The scientific visitors such as Koenig and Bernoulli
sometimes stayed for weeks or months at a
time. Voltaire was pleased that crisp,
Newtonian science was gaining ground
through their efforts. But when he and du
Châtelet engaged in their teasing, their mock
battling, it wasn't the case of a worldly, widely
read man deciding when to let his young lover
win. Du Châtelet was the real investigator of
the physical world, and the one who decided
that there was one key question that had to be
turned to now: What is energy?

$E=mc^2$ · $E=mc^2$ · $E=mc^2$

She knew that most people felt energy was already sufficiently well understood. Voltaire
had covered the seemingly ordained truths in
his own popularizations of Newton: the central factor to look for when you're analyzing
how objects make contact is simply the prod-
uct of their mass times their velocity, or their
mv[1]. If a 5-pound ball is going 10 mph, it has
50 units of energy.

But du Châtelet knew that there had once
been a famous competing view to Newton's,
due to Gottfried Leibniz, the great German
diplomat and natural philosopher. For Leibniz,

the important factor to focus on was mv^2. If a 5-pound ball is going at 10 mph, it has 5 times 10^2, or 500 units of energy.

Which view was true? It might seem a mere quarreling over definitions, but there was something deeper going on behind it. We're used to science being separated from religion, but in the seventeenth and eighteenth centuries it wasn't.

Newton felt that highlighting where mv^1 occurs would prove that God had to exist. If two identical beer wagons crash head-on, there's an almighty bang, and possibly some grinding as their bumpers crumple into each other, but then there's stillness. Right before they hit there was a lot of mv^1 in the universe: the two speeding carts were each loaded with the stuff. One cart had been going full speed due east, for example; the other had been going full speed due west. After they hit, though, and had become stationery chunks of wood and metal, the two separate parts of the v^1 were gone. The "going due east" had exactly canceled out the "going due west."

In Newton's view, this meant that all the energy the carts had once possessed had now vanished. A hole had been created, leading out from our visible universe. Since collisions like this happen all the time, if we live within a great, coglike clockwork, that clock would always need winding. But look around you. We don't find that as the years pass, fewer and fewer objects are able to move. That's the proof. The fact that the universe continues operating was, in Newton's view, a sign that God's reassuring hand was reaching in, to nurture us and to support us; to supply all the motive forces we otherwise lost.

For Voltaire that was enough. Newton had spoken, and who was he to argue with Newton, and anyway it seemed such a magnificent vision—and it was backed by such distressingly complicated geometry and calculus—that it was wisest just to nod in confirmation and accept it. But du Châtelet spent a long time in her room with the Watteau paintings, then at the candle-edged writing table, working through Leibniz's contrary arguments for herself.

Along with various abstract geometric arguments, Leibniz had also focused on the way that Newton's approach left gaps in the world. Diplomats can be sarcastic. He wrote: "According to Newton's doctrine, God Almighty wants to wind up his watch from time to time: otherwise it would cease to move. He had not, it seems, sufficient foresight to make it a perpetual motion."

It turned out that concentrating on measurements of energy as being mv^2 avoided this problem. The mv^2 of a cart going due west might be, say, 100 units of energy and the mv^2 of a second cart going on a collision course due east might be another 100 units. For Newton the two hits canceled each other out, but for Leibniz they added up. If I'm roaring in with 100 units of energy, and you're roaring at me with 100 units of energy, when we hit, the full 200 units of energy we together carry will remain busily in existence, sending metal parts bouncing and rebounding, heating up the wagon wheels, generally creating an ongoing, reverberating jangle.

In Leibniz's view, nothing is lost. The world runs itself; there are no holes or sluicegates where causality and energy rushes away, so that only God would be able to pour them back in. We're alone. God might have been needed at the very beginning, but no longer.

Du Châtelet found some attraction in this view, but also recognized why it had languished in the decades since Leibniz had proposed it. This view was too vague; matching Leibniz's personal biases, but without enough objective proof. It was also, as Voltaire got great satisfaction showing in his novel *Candide,* a strangely passive view; suggesting that no fundamental improvements to our worldly condition could be made.

Du Châtelet was known for being burstingly quick in conversation, but at Versailles that had been because she was surrounded by fools, while at Cirey that was the only way to get a word in with Voltaire. When it came to her original work, she could be much more methodical and took her time. After going through the first arguments by Leibniz, and then the standard critiques against them, she—and various specialists she brought in to help—didn't leave it there, but started looking wider, for some practical evidence that would help her make a choice. To Voltaire she was clearly "wasting" her time, but for du Châtelet it was one of the peak moments of her life: the research machine she had established at Cirey was finally being used to its full capacity.

$E=mc^2$ · $E=mc^2$ · $E=mc^2$

She and her colleagues found the decisive evidence in the recent experiments of Willem 'sGravesande, a Dutch researcher who'd been letting weights plummet onto a soft clay floor. If the simple Emv^1 was true, then a weight going twice as fast as an earlier one would sink in twice as deeply. One going three times as fast would sink three times as deep. But that's not what 'sGravesande found. If a small brass sphere was sent down twice as fast as before, it pushed *four* times as far into the clay. If it was flung down three times as fast, it sunk *nine* times as far into the clay.

Which is just what thinking of Emv^2 would predict. Two squared is four. Three squared is nine. The equation's operation really did seem, in some strange way, fundamental to nature.

'sGravesande had a solid result but wasn't enough of a theoretician to put it all together. Leibniz was a top theoretician but had lacked this detailed experimental finding—his opting for mv^2 had been a bit of a guess. Du Châtelet's work on this topic bridged the gap. She deepened Leibniz's theory, and then embedded the Dutch results within it. Now, finally, there was a strong justification for viewing mv^2 as a fruitful definition of energy.

Her publications had a great effect. Du Châtelet had always been a clear writer, and it helped that Cirey was looked up to as one of the few truly independent research centers. Most English-speaking scientists automatically took Newton's side, while German-speaking ones tended to be just as dogmatically for Leibniz. France had always been the crucial swing vote in the middle. Du Châtelet's voice was key in finally tilting the debate.

After publishing her work she paused to take care of her family's finances and to consider what research topic to do next. There were travels with Voltaire, and she was amused that the new generation of courtiers at Versailles had no idea that she was one of the leading interpreters of modern physics in Europe, or that in her spare time she had published original translations of Aristotle and Virgil. Occasionally it would slip, when she did a burst of probability calculations for the gaming table.

Time passed, and they went back to Cirey. the lime trees were growing ("in this, our delightful retreat," as she wrote), and she had even let Voltaire have his vegetable garden. And then, as she hurriedly wrote in a letter to a friend

3 April, 1749

Chateau de Cirey

"I am pregnant and you can imagine ... how much I fear for my health, even for my life ... giving birth at the age of forty."

It was the one thing she couldn't control. She'd had children shortly after her marriage, but she had been twenty years younger, and even then it had been dangerous. Being this much older, survival was not very likely. Doctors of the time had no awareness that they should wash their hands or instruments. There were no antibiotics to control the inevitable infection; nothing like today's drugs control uterine bleeding. She didn't rage at the clear incompetence of her era's doctors; she just said to Voltaire that it was sad leaving before she was ready. The length of time before her was very clear: the labor was expected in September. She'd always worked long hours; now she sped up, the candles at the desk where she wrote sometimes burning till dawn.

On September 1, 1749, she wrote to the director of the king's library, stating that he would find in the accompanying package the now complete draft of a major commentary she was doing on Newton. Three days later, the birth began; she survived that, but infection set in, and within a week she died.

Voltaire was beside himself: "I have lost the half of myself—a soul for which mine was made."

$$E=mc^2 \cdot E=mc^2 \cdot E=mc^2$$

In time the focus on energy as being proportional to mv^2 began to seem second nature to physicists. Voltaire's polemical skills, passing on the legacy of his lover, helped give it an even stronger boost. In the following century,

Faraday and others used mv²—this quantity that might transform but never totally disappeared—as they built up their visions of the conservation of all energy. Du Châtelet's analysis and writing had been an indispensable step, though in time her role came to be forgotten; partly because each new generation of scientists tends to be generally neglectful of their past; partly, perhaps, because it was unsettling to hear that a woman could have directed such a large research effort and helped shape the course of subsequent thought.

The big question, though, is why. Why is squaring the velocity of what you measure such an accurate way to describe what happens in nature?

One reason is that the very geometry of our world often produces squared numbers. When you move twice as close toward a reading lamp, the light on the page you're reading doesn't simply get twice as strong. Just as with the 'sGravesande experiment, the light's intensity increases *four* times.

When you are at the outer distance, the light from the lamp is spread over the area shown by the patch on the big sphere. When you go closer, that same amount of light gets concentrated on a much smaller area.

The interesting thing is that almost *anything* that steadily accumulates will turn out to grow in terms of simple squared numbers. If you accelerate on a road from 20 mph to 80 mph, your speed has gone up by four times. But it won't take merely four times as long to stop if you apply brakes and they lock. Your accumulated energy will have gone up by the square of four, which is sixteen times, and so that's how much longer your skid will be.

Imagine that skid hooked up to some sort of energy collector. A car that's racing along at four times another one's speed, really will generate—really does carry along—sixteen times as much energy. If someone tried to measure energy as simply equal to mv^1, they'd miss all this. Only by concentrating on mv^2 do these important aspects come out.

Over time, physicists became used to multiplying an object's mass by the square of its velocity (mv^2) to come up with such a useful indicator of its energy. If the velocity of a ball or rock was 100 mph, then they knew that the energy it carried would be proportional to its mass times 100 squared. If the velocity is raised as high as it could go, to 670 million mph, it's almost as if the ultimate energy an object will contain should be revealed when you look at its mass times c squared, or its mc^2. This isn't a proof, of course, but it seemed so natural, so "fitting," that when the expression mc^2 did suddenly appear within Einstein's more detailed calculations, it helped make more plausible his startling conclusion that the seemingly separate domains of energy and mass could be connected, and that the symbol "c"—the speed of light—was the bridge. (For the reader interested in Einstein's actual derivations, my web site dbodanis.com goes through some of his reasoning.)

The c^2 is crucial in saying how this link operates. If our universe were created differently—if c^2 were a low value—then a small amount of mass would only be transformed into an equally small puff of energy. But in our universe, and viewed from the small, ponderously rotating planet to which we're consigned, c^2 is a huge number. In units of mph, c is 670 million, and so c^2 is 448,900,000,000,000,000. Visualize the equals sign in the equation as a tunnel or bridge. A very little mass gets enormously magnified whenever it travels through the equation and emerges on the side of energy.

This means that mass is simply the ultimate type of condensed or concentrated energy. Energy is the reverse: it is what billows out as an alternate form of mass under the right circumstances. As an analogy, think of the way that a few wooden twigs going up in flames can produce a great volume of billowing smoke. To someone who'd never seen fire, it would be startling that all that smoke was "waiting" inside the wood. The equation shows that any form of mass can, in theory, be manipulated to expand outward in an analogous way. It also says this will happen far more powerfully than what you would get by simple chemical burning—there is a much greater "expansion." That enormous conversion factor of 448,900,000,000,000,000 is how much any mass gets magnified, if it's ever fully sent across the "=" of the equation.

Part Three . . .

. . . The Early Years

CHAPTER SEVEN

Einstein and the Equation

When Einstein published $E=mc^2$ in 1905, the equation was at first almost entirely ignored. It simply did not fit in with what most other scientists were doing. The great insights from Faraday and Lavoisier and all the rest were available, but no one else was putting them together this way—hardly anyone even had a hint that one could try.

The world's dominant industries were steel and railways and dyes and agriculture, and that's what ordinary researchers concentrated on. A few universities had specialized labs for more theoretical work, but much of that was in areas what wouldn't have been too surprising to Newton over two centuries before: there were treatises on conventional optics, and sound, and elasticity. There was a little fresh work, on the new and puzzling radio waves, and in areas related to radioactivity, but Einstein was mostly on his own.

We can date to within a month or so the moment when he first saw that E would equal mc². Einstein finished writing his initial paper on relativity by late June 1905, and had the addendum with the equation ready for printing in September, so he probably first realized it some time in July or August. It would likely have been either on one of his walks, or at home after his day job at the patent office. Often his infant son, Hans Albert, was around when he worked, but that wouldn't have been a problem. Visitors recount Einstein contentedly working in the living room of his small apartment, while rocking his one-year-old's bassinet with his free hand, humming or singing to him as needed.

What guided Einstein was that, in his mid-twenties, he found the unknown intriguing. He felt compelled to comprehend what might have been intended for our universe by The Old One (as he referred to his notion of God).

"We are in the position," Einstein explained later, "of a little child entering a huge library, whose walls are covered to the ceiling with books in many different languages. The child knows that someone must have written those books. It does not know who or how. It does not understand the languages in which they are written. The child notes a definite plan in the arrangement of the books, a mysterious order, which it does not comprehend but only dimly suspects."

When the chance came to reach through the gloom, and pluck out The Old One's book that had the shimmering equation $E=mc^2$ written on its pages, Einstein had been willing to take it.

$E=mc^2$ · $E=mc^2$ · $E=mc^2$

The reasoning Einstein used to come up with his extraordinary observation—that mass and energy are done—had begun with the seemingly irrelevant observation that no one could ever catch up with light. But that led, as the space shuttle example suggested, to the insight that energy pouring into a moving object could end up making an outside observer see its mass swell. The argument could also apply in reverse: under the right circumstances an object should be able to pour out energy, generating it from its own mass.

Starting in the 1890s, a few years before Einstein wrote out his equation, a number of investigators had actually seen hints of how this might occur. Several metal-streaked pebbles that had been brought back from the Congo and Czechoslovakia and other places were found, in laboratories in Paris and Munich and Montreal, to be spraying out some sort of mysterious energy beams. If the pebbles were used up as they did this, it wouldn't have been too surprising—one could think that the process was some sort of ordinary burning. But by the best measurements of the time, the energy beams seemed to be pouring out without the pebbles changing in any way.

Marie Curie was one of their first investigators, and indeed in 1898 coined the word *radioactivity* for this active spurting out of radiation. Yet even she, at first, had no understanding that these metals achieved their power by sucking immeasurably tiny portions of their mass out of existence, and switching that mass into the greatly magnified form of sprayed energy. The amounts seemed beyond credibility: a palm-sized chunk of these ores could spray out over 200,000,000,000,000 high-speed alpha particles every second, and repeat this for hours and weeks and months, without any loss of weight than anyone could measure.

Later, after Einstein was famous, he met Curie several times, but he never understood her—after one hiking trip he described her as being cold as a herring and constantly complaining. In fact, she had a passionate nature, was deeply in love with an elegant French scientist who was married to someone else. The reason she complained on the hiking trip was that she was slowly dying of cancer. Radium was one of these scarcely understood new metals, and Curie had been working with it for years.

The minute traces of radium powder, which she had carried unknowingly on her blouse and hands as she walked across the muddy cobblestones of 1890s Paris and later, had been pouring out energy in accord with the then-unsuspected equation, barely shrinking at all, for thousands of years. They had been spray-releasing part of themselves without getting used up back when they were deep underground in the Belgian mines in the Congo; they continued through her years of experiments, ultimately giving her this killing cancer. More than seventy years later, the dust was still "alive," squirting out poisonous radiation onto archivists who were examining her office ledger, and even the cookbooks at her home.

The amount of dust Curie had scattered was measured in millionths of an ounce. But that had been enough, in accord with Einstein's equation, for the radioactive dust to slam into the DNA in her bones, producing the leukemia of which she died; to slam upward, only a fraction more feebly, into the detecting Geiger counters of these startled archivists so many decades later.

Einstein's equation showed how large the result could be. To work it out for any chunk of mass, take the speed of light and square that to get a larger number. Then, multiply that by the amount of mass you're looking at, and that's how much energy, exactly, the mass will be able to pour out.

It's easy to miss how powerful that idea is. For $E=mc^2$ says nothing about what sort of mass can fit into the equation! Under the proper circumstances, *any* substance can have its mass exploded outward as energy. This is the power that's around us, encased within the most ordinary rocks and plants and streams. A single page of this book, weighing only a few grams, seems to be just an innocuous, stable mix of cellulose fibers and ink. But if that ink and cellulose could ever be shifted into the form of pure energy, there would be a roaring eruption, greater than that of a large power station exploding. It's easier to access that power in uranium than in ordinary paper—as we'll see later—but that's simply a limitation of our current technology.

The greater the mass being transformed, the more fearsome the power release. Put a single pound of mass into the "m" slot, and after multiplying by the vast 448,900,000,000,000,000 value of c^2, the equation promises that, in principle, you could get up to 15 billion kilowatt hours of energy. This is greater than all the power stations on Earth. That's how a small bomb—with a core small enough to fit in your cupped hands—could heave out enough energy to rip open streets and buried fuel lines; to shatter street after street of brick buildings; to tear open the bodies of tens of thousands of soldiers and children and teachers and bus drivers.

A uranium bomb works when less than 1 percent of the mass inside it gets turned into energy. An even larger amount of matter, compressed into a floating star, can warm a planet for billions of years, just by seemingly squeezing part of itself out of existence, and turning those fragments of once-solid matter into glowing energy.

$$E=mc^2 \quad E=mc^2 \quad E=mc^2$$

In 1905, when Einstein first wrote out his equation, he was so isolated that he prepared the main relativity article without footnotes. That's almost unheard of in science. The one acknowledgment Einstein did put in was to his loyal friend Michele Besso, a thirty-something mechanical engineer, working at the patent office, who happened to be the author's friend. Even in 1905 physicists complained of being overburdened. Einstein's articles appeared in a distinguished journal—he'd been keen enough on his career to stay connected by submitting review articles—but one after another, the physicists turning through the journal either skimmed or just ignored this exceptional misfit of an article.

At one point Einstein tried applying for a junior teaching position at the university in Bern, as a way out of the patent office. He sent off the relativity article he was so proud of, along with others he'd written. He was rejected. A little later he applied to a high school, again offering his services as a teacher. The equation was sealed in the envelope with the rest of his application forms. There were twenty-one applicants, and three got called in for interviews. Einstein wasn't one of them.

In time a few scientists did begin to hear of his work, and then jealousy set in. Henri Poincaré was one of the glories of Third Republic France, and, along with David Hilbert in Germany, one of the greatest mathematicians in the world. As a young man Poincaré had written up the first ideas behind what later became chaos theory; as a student, the story goes, he'd once seen an elderly woman on a street corner knitting, and then, thinking about the geometry of her knitting needles as he walked along the street, he'd hurried back and told her that there was another way she could have done it: he'd independently come up with purling.

By now, though, he was in his fifties, and although he could still get some fresh ideas, he increasingly didn't have the energy to develop them. Or maybe it was more than that. Middle-aged scientists often say that the problem isn't a lack of memory, or the ability to think quickly. It's more a fearfulness at stepping into the unknown. For Poincaré had once had the chance of coming close to what Einstein was doing.

In 1904 he'd been in the large group of disoriented European intellectuals invited to the World's Fair being held in St. Louis. (Max Weber, the German sociologist, was also there, and he was so startled by the raw energy he saw in America—he described Chicago as being "like a man whose skin has been peeled off"— that it helped jolt him out of a depression he'd been suffering for years.) At the fair, Poincaré had actually given a lecture on what he'd labeled a "theory of relativity," but that name is misleading: it only skirted around the edges of what Einstein would soon achieve. Possibly if Poincaré had been younger he could have pushed it through to come up with the full results Einstein did the next year, including the striking equation. But after that lecture, and then the exhausting schedule his St. Louis hosts had for him, the elderly mathematician let it slide. The fact that so many French scientists had turned away from Lavoisier's hands-on approach and instead insisted on a sterile overabstraction only made it harder for Poincaré to be immersed in practical physics.

By 1906, realizing that this young man in Switzerland had opened up an immense field, Poincaré reacted with the coldest of sulks. Instead of looking closer at this equation, which he could have considered a stepchild, and bringing it in to his Paris work team for further development, he kept a severe distance; never speaking of it; seldom mentioning Einstein's name.

Other contemporaries did examine Einstein's work more closely, but tended to miss, at first, such key points as why Einstein selected "c" as being so central. They could understand if relativity and the equation had come from some fresh experimental results; if Einstein had built some new-style apparatus in a laboratory to look more closely at what Marie Curie or others were finding, and so had discoveries which no one else did. But what they could not grasp was that he didn't *have* any labs. The "latest findings" he worked with came from scientists who'd died decades or even centuries before. But that didn't matter. Einstein hadn't come up with his ideas by patiently putting together a range of new results. Instead, as we saw, he just spent a long time "dreamily" thinking about light and speed and what was logically possible in our universe and what wasn't. But it only seemed "dreamy" to outsiders who didn't understand him. It became one of the major intellectual achievements of all time.

$$E=mc^2 \cdot E=mc^2 \cdot E=mc^2$$

For centuries after the birth of mathematically guided science around the seventeenth century, humans thought that they had the main lines of the world described, and that although there were further details to work out, the "commonsense" properties of the world around us could be taken for granted. We lived in a world where objects kept a constant mass as they moved around; where time advanced smoothly, and everyone could agree about where we were in its flow.

Einstein saw that the universe was different from what everyone had thought. It was, he realized, as if God had restricted us to a small playpen—the surface of the Earth—and had even let us think that what we observed from it was all that really occurred. Yet all the while, stretching further out—around us all the time if we were able to see it—was a further domain, where our intuition no longer applied. Only pure thought would allow us to see what happened there.

The fact of energy and mass being interchangeable, as shown in E=mc², is only one of these fuller consequences. There are others as well, and to recognize them, it helps to imagine a world where instead of the uppermost speed limit being the speed of light at 670 million, it is instead an easy 30 mph. What does Einstein's 1905 relativity paper say we'll see?

The first striking thing we would see if we entered that world follows from the space shuttle example. Cars would have their ordinary weight when they were waiting patiently at a red light, but once the light changed to green, they would bulk up in mass as they got faster. It would happen to pedestrians and joggers and bicyclists and indeed everything that moved. A schoolchild, who might weigh 100 pounds on her bicycle when waiting at a corner, would bulk up to 200 pounds once she had pedaled up to 20 mph. If she was fast, or had a downhill slope to help and got up to 29 mph or more, she'd soon have a mass of 3,000 pounds or more. Her bicycle would swell up just as much. As soon as she stopped pedaling both she and her bicycle would immediately come down to their original, static weight.

At the same time, cars, bicycles, and even pedestrians would undergo another change. As they moved forward, someone watching from very close by would see them begin to rotate, so that a 12-foot car would appear to be lifted up slightly at one end once it got going fast enough, and if it reached 29.99 mph, it might be almost balanced on its front fender as it skidded along. The driver and passengers inside would have rotated just as much—and again, as soon as they stopped, they would settle back like a teeter-totter ride ending, and return to their normal position.

As the cars hurried by, we'd not only see them as getting more massive and swiveling, we'd also notice that time seemed to be slowing down inside of them. If the driver reached to turn on the CD player, we'd see his hand move in extreme slow motion. Once the player was on and the sound was coming out, we'd hear the sound waves transmitted with painful slowness, transforming even early Michael Jackson warbles into heavy, dirgelike chants.

In this view of the universe there is no "true" perspective—some sort of traffic helicopter hovering above this odd city—from which one can assert that, yes, the cars are undergoing some strange changes, but the bystanders who aren't moving are unchanged and clearly "normal." For why should the bystanders have some favored status, while it's only the moving cars that are changing? In fact, the drivers of the cars, or the schoolgirl on her bicycle, will have no sensation that they're changing. The bicyclist will look around her, and see that her handlebars and her body and her backpack haven't become heavier. Rather, to her it's the people left behind who will seem weird. *They*'ll be the ones whose mass has swelled.

The passengers in the car will agree. Their CD player is fine, they'll say, and the young Michael Jackson is warbling along as quickly as ever. It's the people outside the car who seem slow, with hotel doormen seeming to lift their arms in laborious heaviness, then puffing out their cheeks like stately deep-sea fish whenever they blow a whistle to hail a taxi.

These effects are summarized in relativity by saying that when someone watches an object recede away from them, that object will be seen to undergo mass dilation, length rotation, and time dilation. The bystanders will see it in the car; the driver of the car, looking back, will see it in the bystanders.

The first time one reads of this, it seems like nonsense. Even Einstein found it hard to accept—as with the inexplicable tension he felt in his evening talk with Michele Besso, on the summer day when he was still trying to work out these relations. But, it's only hard to accept because we never actually interact with each other at speeds close to the 670 million mph of light (and the effects are too slight to notice at our ordinary speeds). Think, for example, of a portable music player at a picnic. To someone standing next to it, it's loud. To someone who walks a few hundred yards away, the music is soft. We accept that there's no answer about how loud it "really" is. But that's simply because we're capable of walking quickly enough to cover that few hundred yards in a brief time. To an ant or some smaller creature, one that took many generations to migrate far enough from the music player to detect any change in its volume, our view— that music can appear to be at different volumes to different observers—would seem crazy.

All these details follow from such simple observations as light's constancy. But there are a number of ordinary objects around us, which always do work at the high speeds where these effects become apparent. The electrons that shoot from the back of traditional TV sets to the screen at the front, for example, travel so fast that they really will seem to us as if they've grown in mass as they travel. Engineers have to take that into account when they design the magnets that focus the screen's image. If they didn't we would see a blur.

The Global Positioning System (GPS) of navigational satellites, which fly overhead and beam down location signals for cars and jets and hikers, are also traveling so fast that from our perspective, time on board them seems to be slow. The circuits in the handheld GPS location devices we use to locate our positions, or in the larger GPS devices that banks use to synchronize payments, are programmed to correct for this—in exact accord with the equations Einstein worked out in 1905.

Einstein never especially liked the label *relativity* for what he'd created. He thought it gave the wrong impression, suggesting that anything goes: that no exact results any longer occur. That's not so. The predictions are precise.

The label is also misleading because all Einstein's equations are cohesive, and exactly linked up. Although each of us might view things in the universe differently, there will be enough synchronization where these different views join to ensure that it all fits. The notions that mass never changes and that time flows at the same rate for everyone made sense when people only noticed the ordinary, slow-moving objects around them. In the true wider universe, however, they're not true—but there are exact laws to explain how they change.

This is an achievement that has occurred very few times in history. Imagine being able to make a shimmering crystalline model, small enough to hold in your closed fist. Now open your hand—and see the entire universe soar out; glowing into full existence. Newton was the first person to have done that, back in the 1600s: conceiving a complete system of the world, that could be described in but a handful of equations, yet also contained the rules for how to move out from the summary and go on to creating the full world. Einstein was the next.

Just to make the bond more impressive, both Einstein and Newton achieved much of their work in impossibly brief periods in their mid-twenties. For Newton, back at his mother's Lincolnshire farm after his university had been closed because of the Plague, there were about eighteen months in which he did fundamental work on developing calculus, conceiving the law of universal gravitation, as well as working on key concepts for a mechanics that would apply throughout the universe. For Einstein, in a period of under eight months in 1905—and while still putting in full days at the patent office, Monday through Saturday—there was his first theory of relativity, and $E=mc^2$, as well as his work that helped lay the path for lasers, computer chips, key aspects of the modern pharmaceutical and bioengineering industry, and all Internet switching devices. He really was, as Newton described himself at that same period, "in the prime of my age for invention." In each area, Einstein pushed beyond what was known; he unified fields that had remained separate, questioning assumptions that everyone until then had simply accepted.

The few researchers around 1905 who had uncovered a small part of what he later deduced had no chance of matching him. Poincaré got closer than almost anyone else, but when it came to breaking our usual assumptions about time's flow or the nature of simultaneity, he backed off, unable to consider the consequences of such a new view.

$E=mc^2$ · $E=mc^2$ · $E=mc^2$

Why was Einstein so much more successful? It's tempting to say it was just a matter of being brighter than everyone else. But several of Einstein's Bern friends were highly intelligent, while someone like Poincaré would have been off the scale on any IQ test. Thorstein Veblen once wrote a curious little essay that I think gets at a deeper reason. Suppose, Veblen began, a young boy learns that everything in the Bible is true. He then goes to a secular high school, or university, and is told that's wrong. "What you learned at your mother's knee is entirely false. What we teach you here, however, will be entirely true." Some students would say, Oh, fine, I'll accept that. But others will be more suspicious. They'd been fooled once before, taking on faith an entire traditional world. They're not going to be fooled again. They would learn what was on offer, but always hold it critically, as just one possibility among others. For Einstein, being Jewish—and even though his immediate family wasn't observant—meant being immersed in a culture with different views about personal responsibility, justice, and belief in authority than the standard German and Swiss consensus.

There's more, though. When Einstein was a little boy, he was fascinated with how magnets worked. But instead of being teased about it by his parents, they accepted his interest. How *did* magnets work? There had to be a reason, and that reason had to be based on another reason, and maybe if you traced it all the way, you'd reach . . . What would you reach?

At one time, in the Einstein household, there had been a very clear answer to what would ultimately be reached. When his grandparents had been growing up, most Jews in Germany were still close to traditional Orthodoxy. It was a world suffused by the Bible, as well as by the crisply rational analysis of the Talmud. What counted was to push through to the very edge of what was knowable, and comprehend the deepest patterns God had decreed for our world. Einstein had gone through an intense religious period when he was approaching his teens, though by the time he was at the Aarau high school that literal belief was gone. Yet the desire to see the deepest underpinnings was still there, as was the trust that you would find something magnificent waiting if you made it that far. There was a waiting "slot": things could be clarified, and in a comprehensible, rational way. At one time the slot had been filled in by religion. It could easily enough be extended now to science. Einstein had great confidence that the answers were waiting to be found.

It also helped that Einstein had the space to explore his ideas. The patent job meant that he didn't have to churn out academic papers ("a temptation to superficiality," Einstein wrote, "which only strong characters can resist"), but rather could work on his ideas for as long as it took. Most of all, his family trusted him, which is a great boost to confidence, and they also encouraged a playful, distancing tone. It's just what's needed for "stepping back" from ordinary assumptions, and imagining such oddities as a space shuttle pushed up against a barrier at the speed of light, or someone chasing toward a skedaddling beam of light.

His sister, Maja, later gave a hint of this gently self-teasing tone. When Einstein got in a temper as a little child, she recounted, he sometimes threw things at her. Once it was a large bowling ball; another time he used a child's hoe "to try to knock a hole" in her head. "This should suffice," she commented, "to show that it takes a sound skull to be the sister of an intellectual." When she described the high school Greek teacher who complained that nothing would ever become of her brother, she added: "And in fact Albert Einstein never did attain a professorship of Greek grammar."

To crank it all forward, there need to be driving tensions, and these Einstein had aplenty. There was the "failure" of being in his mid-twenties, isolated from other serious scientists, when university friends were already making careers for themselves. There was also thunderous guilt from seeing the difficulties his father was having in his own business career. Einstein had grown up with his father fairly prosperous in the electrical contracting business in Munich, but when Einstein was a teenager, partly because key contracts stopped being given to Jewish firms, and his father had moved the family to Italy to set up again. In the move, and in a series of near-successes that never quite made it, his father was exhausted paying back loans to a brother-in-law, the constantly nagging Uncle Rudolf "The Rich" (as Einstein mockingly called him). It wrecked his father's health; yet through it all the family had insisted on finding the money to pay for Einstein to study. ("He is oppressed by the thought that he is a burden on us, people of modest means," as his father had remarked in the 1901 letter.) There was a huge obligation for Einstein to show he had been worth it after that.

$E=mc^2$ · $E=mc^2$ · $E=mc^2$

Eventually a few other physicists did begin to pay attention to Einstein, sometimes visiting Bern to talk over the equation and other results. It was just what Einstein and Besso had hoped for, but it also meant that they started being pulled apart. For Einstein was gradually going beyond the ideas his best friend could follow. Although Besso was bright, he'd chosen a life in industry. ("I prodded him very much to become a university teacher, but I doubt . . . he'll do it. He simply doesn't want to.") Besso couldn't follow the next level.

Besso adored his younger friend, and had gone out of his way to help him back when Einstein was still a student. He even tried, hard, in their evenings sharing Gruyere and sausages and tea, to keep up with the further ideas Einstein was seeing now. Einstein himself was kind about the growing distance from his friends. He never declared to Besso that he was no longer interested in him. They continued country walks, stops for a drink, musical evenings, and practical jokes with the others. But it's a bit like two old school friends breaking off once both have started moving separate ways at university, or in their first jobs afterward. Each one would really like things not to be like that, but everything they care about now is pulling them apart. They can talk about the old days when they're together, but the enthusiasm is forced, even though neither of them wants to admit it.

A similar distancing happened with Einstein's wife, Mileva. She'd been a physics student with him, and very bright. Men in the sciences rarely marry fellow specialists—how many are there?—and Einstein was almost smug to his college friends about how lucky he'd been. His first letters to her had started neutrally:

> Zurich, Wednesday 16 February 1898
>
> I have to tell you what material we covered.... Hurwitz lectured on differential equations (exclusive of partial ones), also on Fourier series....

But the relationship developed, as extracts from a series of letters written in August and September 1900 show:

> Once again a few lazy and dull days flitted past my sleepy eyes, you know, such days on which one gets up late because one cannot think of anything proper to do, then goes out until the room has been made up....Then one hangs around and looks halfheartedly forward to the meal....
>
> However things turn out, we are getting the most delightful life in the world. Beautiful work, and together....
>
> Be cheerful, dear sweetheart. Kissing you tenderly, your
>
> Albert

The life they shared started out happily. His wife wasn't going to be at his level, but she really was a good student. (On the university final exams where he scored 4.96, she came close, with a 4.0. She certainly could have followed his work. (The myth that she had been responsible for his key work stems from nationalist Serb propaganda in the 1960s, as her family had originally been from near Belgrade.) But once their children came, and with their income so low that they only had part-time help, all the traditional sexism took over. When educated friends came to visit, his wife would try to join in, but it was not easy with an attention-frantic three-year-old son on your lap. You want to stay a part of the conversation, but after too many interruptions for getting toys and drawing pictures and picking up spilled food, the guests no longer stop their talk to recap things and bring you in. You're left out.

Einstein finally left the patent office—though even when he did, in 1909, his chief was mystified as to why this young man was willing to turn his back on such a good career. He was finally offered a position in the Swiss university system, and after a stint in Prague—where he played music and engaged in discussions at a salon that sometimes included a shy young man named Franz Kafka—Einstein ended up as a professor in Berlin. His success had now isolated him nearly complete from his Bern friends. He was separated from his wife, and only occasionally got to see his adored two children.

By that time, Einstein himself was taking his personal work in a different direction. The equation $E=mc^2$ was just one small part of the special theory of relativity. By 1915, he'd perfected an even grander theory, so powerful that the special theory was just one small part of that. (The Epilogue gives some highlights of that 1915 work—"compared with this problem, the original theory of relativity is child's play.") He would be involved with the equation only once more, briefly, when he was a much older man.

At this point, there's a major shift in the story. The equation's first theoretical development was over; Einstein's personal contribution fades away. Europe's physicists accepted that $E=mc^2$ was true: that, in principle, matter could be transformed so that the frozen energy it was composed of could be let out. But no one knew how to actually get that to happen.

There was one hint. It came in the strange objects that Marie Curie and others were investigating: the dense metals of radium, and uranium, and other substances, which were somehow able to pour out energy week after week, month after month; never using up whatever "hidden" source of supply they contained inside.

A number of laboratories began to study how that might be happening. But to see what mechanisms were creating these great out-wellings of energy, it wouldn't be enough to continue looking at the surface of things, simply measuring the weight or color or surface chemical properties of the mysteriously warm radium or uranium.

Instead, the researchers would have to go within, deep into the very heart of these substances. That, ultimately, would show how the energy that $E=mc^2$ promised could be accessed. But what would they find, as they tried to peer into the smallest, inner structures within ordinary matter?

CHAPTER EIGHT

Into the Atom

University students in 1900 were taught that ordinary matter—bricks and steel and uranium and everything else—was made of smaller particles, called atoms. But no one knew for sure what these atoms were made of. One common view was that they were something like tough and shiny ball bearings: mighty glowing entities that no one could see inside. It was only with the research of Ernest Rutherford, a great, booming bear of a man working at England's Manchester University, in 1909, that anyone got a clearer view.

Rutherford was at Manchester, rather than at Oxford or Cambridge, not just because he was from rural New Zealand, and spoke with a common man's accent. If a research assistant was self-effacing enough, that could be overlooked. The problem rather was that when

Rutherford had been a student at Cambridge he had refused to show proper deference to his superiors. He'd even suggested creating a joint-venture business to earn money from one of his inventions—and that was a mortal sin. Yet the reason he became the scientist who got the first clear glimpse of the inside of atoms was, to a large extent, because his heightened awareness of discrimination made him the kindest leader of men. The bluff booming exterior was just window dressing. He was good in nurturing skilled assistants, and the 1909 experiment was monitored by a young man who would end up perfecting a most useful mobile radiation detection unit, of Rutherford's suggested design: the audibly clicking counter was to be Hans Geiger's claim to fame.

Their finding is so widely taught in schools today that it's hard to get back to the time when it was still surprising. What Rutherford realized was that these solid, impregnable atoms were almost entirely empty. Imagine that a meteor plummets into the Atlantic Ocean, but instead of staying down there, ultimately plonking against the seabed, we hear a great roaring, and see it come hurtling back out. Think how hard it would be to break through our preconceptions, and realize that the only way to explain it was that under the surface of the Atlantic there really wasn't smooth water all the way down. Rather, the analogy with what Rutherford had to deduce would be that the Atlantic's surface was just a thin liquid-rubbery film, and underneath it, where we had always thought there were deep waves and currents and tons of water, there was . . . Nothing.

It was all empty air, and a camera down there would show the arriving meteor, once it pierced the outer film, falling through empty space. Only at the very center, down on the sea floor, was there some powerful device, extremely compact, that could grab an incoming meteor, and send it hurtling up through the atmosphere, and back into outer space. The equivalent inside an atom is the atom's nucleus, lost far in the center. Only up near the outer surface of an atom are there the flurries of electrons that are involved in ordinary reactions, such as burning a piece of wood in a fire. But they're far from the central nucleus, which is shimmering deep below, within all the empty space.

If atoms were like little ball bearings, then Rutherford had found that these ball bearings were almost entirely hollow. There was just a tiny speck right at the center, called the nucleus. It was a disconcerting finding—the atoms we're composed of are mostly just empty space!—but by itself that still wouldn't have let anyone see how $E=mc^2$ could apply. The "solid" electrons up on the outer surfaces of the atom weren't going to pop out of their material existence and turn into exploding clouds of energy.

It was pretty clear that the nucleus was where scientists would have to turn next. There was a lot of electricity in the atom, and while half of it was spread diffusely, in the far-flung orbits of those electrons, the other half of it was crammed into the ultradense nucleus at the center. There was no known way to keep so much electricity concentrated in that small a volume. Yet something down there, in that nucleus, was able to squeeze down all that electricity, and hold it in a tight grasp, and keep it from squirmingly escaping. That must be where the storehouse—the hidden energy—that Einstein's equation hinted at could reside. There were positively charged particles—what we call the proton—in there, but no one could make out any greater detail.

An assistant of Rutherford's, James Chadwick, finally got an important better view, in 1932, when he detected yet another item locked inside the nucleus. This was the neutron, which got its name because although it roughly resembled the proton in size, it was electrically entirely neutral. It had taken

Chadwick more than fifteen years to identify it. At one point students had put on a play about his quest for this particle that had so few properties it might as well, they teased, be called the "Fewtron." But if you've spent years putting up with Rutherford's booming impatience, you can handle students having their fun. Although Chadwick was a quiet man, he was pretty determined about what he would do.

Chadwick had originally been a slum boy from the Manchester streets, and his professional career had almost been destroyed just as it was about to begin. As a new postdoc under Rutherford, Chadwick had gone to Berlin, to study in the labs of the returned Hans Geiger. When World War I began, he meekly followed the advice of the local Thomas Cook's office that there was no reason to hasten to leave. As a result, he ended up spending four years as a POW, in the converted stables of a cold and windy Potsdam racecourse. He tried doing as much research as he could there, and even managed to get radioactive supplies. An enterprising firm, the Berlin Auer company, had extra thorium, and was marketing it to the German public in toothpaste as a way to make your teeth glow white. Chadwick simply ordered this miracle tooth whitener from the guards, then used it for his experiments. But he had such poor equipment that his tests never came to much. He was falling behind, and when he got back to England at the war's end in late 1918, barely managed to get back on track. Never again would he meekly follow anyone's advice.

In theory Chadwick's 1932 discovery of the neutron should have led immediately to further discoveries. A number of radioactive substances release neutrons, and those could be aimed like a submicroscope machine gun at waiting atoms. Because they were neutral, they wouldn't be bothered by the negatively charged electrons at the surface of the target atoms. When they reached the nucleus at the center, they shouldn't be bothered by any charges down there either. They'd be able to slip right in. Maybe you could use them as probes to see what was happening in there.

To Chadwick's disappointment, though, he could never get that to happen. The harder he blasted neutrons in at an atom, the less success he had in getting any of them to slip into the nucleus at the center. Only in 1934 did yet another researcher find a way around that problem, and manage to get neutrons to enter easily inside a target nucleus, to better see that nucleus's structure. And he wasn't working in an even more sophisticated research lab, but in the last place one might have expected.

The city of Rome, where Enrico Fermi lived, had memories of grandeur, but in the long decades leading up to the 1930s, it had steadily been left further and further behind the rest of Europe. The lab that the government gave Fermi, who was respected as one of Europe's leading physicists, was on an out-of-the-way street, in a quiet gardened park. There were tiled ceilings, and cool marble shelves, and a goldfish pond under the shady almond trees out back. For someone wanting to make a break from the mainstream European consensus, it was ideal.

What Fermi found in this gentle seclusion was that other research teams had been wrong to focus on blasting neutrons at higher and higher power to get them to enter the tiny nucleus of an atom. Spraying fast neutrons directly at the great empty spaces inside a target atom meant that most of the neutrons simply raced right through. Only if the neutrons were *slowed,* so that they almost dawdled in their flight toward a distant nucleus, would they have a good chance of slipping inside. Slowed neutrons acted like sticky bullets. The reason they stuck so well to nuclei was that they became "spread out" in their relatively slow flight. Even if their main body missed the nucleus, the spread-out portions were still likely to connect.

On the afternoon when Fermi realized slow neutrons could do this, his assistants lugged up buckets of water from the goldfish pond out back. They sprayed fast neutrons from their usual radioactive source into the water. The water molecules were of a size that made the incoming fast neutrons rebound back and forth till they slowed down. When the neutrons finally emerged, they were traveling slowly enough to slip regularly into any target nuclei ahead of them.

With Fermi's trick, scientists now had a probe that could get into the nucleus. But even that didn't make things entirely clear. For what was happening when the slowed neutrons entered? The full power that Einstein's equation spoke about still wasn't coming out. At most you got slightly changed forms of ordinary nuclei, which leaked out only a gentle sort of energy. It was useful for tracers that could be swallowed and then tracked to see what was going on inside the body. One of the first researchers to use similar tracers, George de Hevesy, employed it, his very first time, to prove that the "fresh" hash his Manchester boardinghouse landlady was serving was not quite as fresh as promised, bur rather was coming back, slopped onto a fresh plate, steadily every day. But the slight energy leakages from elements that could be safely swallowed were not what the massive c^2 in the equation promised.

Somehow there had to be a further explanation; some further level of detail that physicists hadn't yet grasped. Atoms weren't solid massy spheres, but rather were almost entirely empty space—like an emptied ocean basin—with just the barest speck of a nucleus down at the center. That was what Rutherford had seen. The nucleus wasn't a simple solid speck either. It contained protons that crackled with positive electric charge, and pebblelike neutrons were packed in along with them. That was clear by 1932. The neutrons could go in and out of that nucleus pretty easily, once you did the unexpected twist of slowing them down when you sent them forward. That was what Fermi saw in 1934. But that's where matters stuck for several years.

CHAPTER NINE

Quiet in the Midday Snow

The solution to what was happening inside the nucleus—and so an unveiling of matter's deeper mechanisms, which would finally allow the energy promised by $E=mc^2$ to emerge—only came in 1938. It was provided by a solitary Austrian woman, sixty years old, stuck on the edge of Europe, in Stockholm; who didn't even speak Swedish.

"I have here . . ." she wrote, "no position that would entitle me to anything. Try to imagine what it would be like if . . . you had a room at an institute that wasn't your own, without any help, without any rights. . . ." It was a dispiriting change, for just a few months earlier, Lise Meitner had been one of Germany's leading scientists—"our Madame Curie," as Einstein put it.

She'd first arrived in Berlin in 1907, an impossibly shy student from Austria. But she'd tried to open up, and soon made friends with one exceptionally good-looking young man at her university, named Otto Hahn. He had an easygoing confidence, a self-teasing Frankfurt accent, and seemed to feel it a personal obligation to put this quiet newcomer at ease.

They were soon sharing a lab in the basement of the Chemistry Department. They were almost exactly the same age, in their late twenties. He persuaded her to hum two-part harmony songs from Brahms with him, despite her off-key voice. When their shared work was going especially well, she wrote, "Hahn would whistle large sections of the Beethoven violin concerto, sometimes purposely changing the rhythm of the last movement just so he could laugh at my protests. . . ." The Physics Institute was nearby, and other young researchers there "often visited us and would occasionally climb in through the window of the carpentry shop, instead of taking the usual way." After working hours, Meitner remained solitary, living in a succession of single rooms, and sitting in the cheapest student seats at concerts she went to by herself. It was only at the lab that she found community.

She was a much better analyst and theoretician than Hahn, but he was bright enough—

and sensible enough—to realize this would only be to his good; he had a history of finding excellent mentors. The first joint discoveries of Meitner and Hahn led to their getting a large lab in the new Kaiser Wilhelm Institutes, on what was then the western outskirts of Berlin. There were rural windmills still within sight; a forest a little farther to the west. They were becoming known as an important and trustworthy research team; they contributed to building up a core of indispensable knowledge about how atoms worked; their findings were soon as necessary to consider as those of Rutherford in England.

Through it all, she and Hahn kept their surface formality, carefully avoiding the informal "Du" form of address. In all her letters he was "Dear Herr Hahn." But there can be a special bond this way; a carefully unstated awareness that such dignified formality is blocking the pair off from any deeper links.

In 1912, after four years of working together, with Meitner now age thirty-four, Hahn married a younger art student. Meitner told everyone that it didn't matter. But although she'd never officially dated Hahn, she never dated anyone else in the years after that. There was another young colleague Meitner had been friendly with, James Franck, and she stayed in touch with him for over half a century, even when he got married, and then later when he was forced out of Germany to distant America. "I've fallen in love with you," Franck teased when they were both in their eighties. "Spät! (Late!)" Lise laughed.

In World War I, Meitner volunteered in hospitals, including some hellish ones near the eastern battlefields, while Hahn was on assignment with the army. The moral dilemmas of his work with poison gas seemed to bother neither of them. She sent letters regularly: lab gossip, and accounts of swimming trips with Hahn's wife, and occasionally the gentlest description of her hospital work. She also had a little time for research: "Dear Herr Hahn! . . . Take a deep breath before you begin reading. . . . I wanted to finish some of the measurements so that I could . . . tell you a variety of delightful things."

Meitner had filled in one of the last gaps left in the periodic table listing all the elements. The work was her own, but she put both their names on it, and insisted to the *Physikalische Zeitschrift* editor that Hahn's name go first. During their wartime separations she tried not to push him for replies, but sometimes she slipped: "Dear Herr Hahn! . . . Be well, and write, at least about radioactivity. I remember a time very long ago when you would once in a while send a line even without radioactivity."

A little after the war they switched to different labs. By the mid-1920s Meitner headed the theoretical physics division within the Kaiser Wilhelm Institutes. She was still shy on the outside, but had become confident in her intellectual work, regularly sitting in the front row with Einstein or the great Max Planck at the most respected theoretical seminars. Hahn was aware he couldn't follow such explorations, and cautiously stuck to more straightforward chemistry. But when Fermi's 1934 advances showed how the neutron might offer an ideal probing tool into the nucleus, Meitner shifted once again, to studies of the nucleus's properties—and this meant she could hire Hahn: chemists were always needed to study the new substances that were being formed.

In 1934 they started working together again, also taking on a recent doctoral student as their assistant, Fritz Strassman. Hitler had come to power in 1933, but although Meitner was Jewish, and so immediately fired from the University of Berlin, she still was an Austrian citizen. The Kaiser Wilhelm Institutes had its own source of funding, and happily continued paying her as a full staff member.

But in 1938, Germany took over Austria, and Meitner became a German citizen by default. The institute might still be able to keep her on, but it would depend a lot on what her colleagues said. An organic chemist named Kurt Hess had long had a small office at the institute. He was a minor researcher, full of envy, and he was one of the first at the institute to become an active Nazi. "The Jewess endangers our institute," he began to whisper, to anyone who would listen. Meitner heard this from one of her ex-students, who had remained loyal. She talked it over with Hahn. Hahn went straight to Heinrich Hörlein, the treasurer of the organization that funded the Kaiser Wilhelm Institute for chemistry.

And Hahn asked Hörlein to get rid of Meitner.

To say that people have been charming, as Hahn had been all his life, is simply to say that they've developed a reflex to do what will put the individuals around them at ease. It says nothing about their having a moral compass deeper than that. Hahn may have been slightly troubled by what he was doing to his old colleague: "Lise was very unhappy now that I had left her in the lurch." But most other German physicists did what the new government wanted them to, and many of Hahn's past students, pro-Nazi, were in positions of power as well. They—more than she—were the people he was increasingly working with now, the ones he needed to please.

He helped her a little bit with the details of leaving, but it's unclear how much Meitner understood in the shock. From her diary: "Hahn says I should not come to the Institute anymore. He has, in essence, thrown me out."

By the time she'd settled in Stockholm, in August 1938, Meitner didn't mention to anyone else what Hahn had done. Instead, almost by reflex, she just remained involved from a distance with the work she had been leading. With Strassman and Hahn's help, she'd been guiding the streams of slowed neutrons into uranium, the heaviest of all naturally occurring elements. Since neutrons slipped into and then stuck within the nuclei they hit, everyone expected that the result would be some new substance, even heavier in weight than the uranium they started with. But try as she and the researchers in Berlin might, they couldn't clearly identify whatever new substances they were creating.

Hahn, as ever, seemed the slowest to grasp what was happening. Meitner met him, in neutral Copenhagen in November, and after he admitted he didn't have a clue, she sent him back with clear instructions for more experiments. He just had to use the top-quality neutron sources and counters and amplifiers she'd assembled, and which were still in place in their lab, right where she'd left them. The mail was so quick between Stockholm and Berlin that she could even talk him through the steps. "Meitner's opinion and judgment carried so much weight with us in Berlin," Strassman recounted later, "that we immediately undertook the necessary . . . experiments." However wounded she was, at least she could continue with the work that had been her focus for years.

She suggested they keep an eye out for variants of radium that might be produced in the long bombardment process that had started with uranium. (Radium is a metal with a nucleus almost as massive as that of uranium. Both are so overstuffed with neutrons that they regularly end up spraying out radiation.) At this stage it was just a hunch, based on similarities between the two metals, and the fact that they were so often found together in mines.

But it meant that the broader effects of $E=mc^2$ were, finally, about to appear.

Monday evening in the lab

Dear Lise!

…There is something about the "radium isotopes" that is so remarkable that for now we are telling only you.… Perhaps you can suggest some fantastic explanation.... If there is anything you could propose that you could publish, then it would still in a way be work by the three of us!

Otto Hahn

They had been using ordinary barium as something of an adhesive in the lab, to gather the fragments of neutron-loaded radium. Once the barium had done that job, it was collected with acids and then rinsed away. The problem now, though, was that Hahn could not get it to separate. Some of the barium that was left always seemed to have tiny bits of something radioactive stuck to it.

He and Strassman were at a loss. "Meitner was the intellectual leader of our team," Strassman explained. But now she wasn't here. Hahn wrote her again, two days later: "You see, you will do a good deed if you can find a way out of this." They could do no more. The strange result—why couldn't they get the radiation away from the simple barium?—would be up to her to try to work out.

It was Christmas by this time, and a couple who knew that Meitner was alone in Stockholm invited her to stay at a hotel in their vacation village of Kungalv, on the west coast of Sweden. A nephew of hers whom she'd always liked, Robert Frisch, was in Copenhagen, and on Meitner's suggestion, the couple invited him too.

Meitner had first really come to know her nephew when he'd been an eager science student in Berlin a decade before. They'd often played piano duets together, even though she had trouble keeping up. (Though they'd have fun, translating *Allegro ma non tanto* as "Fast, but not auntie.")

Now Robert was a grown man, and a promising physicist, working at Niels Bohr's institute in Denmark. The first night, arriving late Christmas Eve, he was in no condition for discussing science. The next morning, when he came down to the ground-floor restaurant at their hotel, he found his aunt puzzling over Hahn's letter. The barium they had added was showing such persistent radioactivity—so much spraying out of energy streams—that she and the researchers in Berlin couldn't help but wonder why. Had it somehow been created that way during the Berlin experiments?

Frisch suggested that it was just a mistake in Hahn's experiment, but his aunt waved that aside. Hahn was no genius, but he was a good chemist. Other labs made mistakes. Not hers. Frisch didn't take much convincing. He knew she was right.

They stayed at the breakfast table while Frisch ate, talking it over. The experiment that Meitner had suggested to the Berlin crew could be explained if the uranium atom had somehow split apart. A barium nucleus is about half the size of a uranium nucleus. What if the barium they were detecting was simply one of the big halves that had resulted? But by everything nuclear physics had been showing—all the work from Rutherford on up—that should be impossible. There are over 200 particles inside a uranium nucleus, all those protons and neutrons. They were stuck together with what was being termed the strong nuclear force, an exceptionally powerful nuclear glue. How could a single incoming neutron break through every one of those bonds, to tear off a huge chunk? You don't throw a simple pebble at a large boulder and expect it to break in half.

They finished breakfast, then went for a walk in the snow. Their hotel wasn't far from a forest. Frisch put on skis, and offered to help his aunt with a pair for herself, but she declined ("Lise Meitner made good her claim," Frisch wrote, "that she could walk just as fast without").

No one had ever chipped off more than a fragment from a nucleus. They were confused.

Even if an incoming neutron had hit some sort of weak point, how could dozens of protons be pulled off in one impact? The nucleus wasn't built like a rocky cliff that could break in half. It was supposed to remain intact for billions of years.

Where could the energy to suddenly tear it apart come from?

Meitner had first met Einstein at a conference in Salzburg, in 1909. They were almost exactly the same age, and Einstein was already famous. At the conference he recapped his 1905 findings. To find that energy could appear out of disappearing mass was "so overwhelmingly new and surprising," Meitner explained decades later, "that to this day I remember the lecture very well."

Now, in the snow with her nephew, she stopped by a tree trunk, and they settled down to work it out. The most recent model of the nucleus was due to Niels Bohr, the kindly soft-spoken Dane who was her nephew's employer. Instead of looking at the nucleus as a rigid metal, some stiffened collection of ball bearings welded tight, Bohr viewed it more as a liquid drop.

A water drop is always on the verge of bursting apart, due to the weight inside it. That near-bursting weight is like the electric forces between the protons in a nucleus. All the protons push against each other. (That's what two positive charges will always do.) But a water drop stays together, most of the time, because it also has a lot of rubbery surface tension on the top. That is like the glue-taut strong force that clusters the protons together, despite all the electricity trying to push them apart.

In a small nucleus, such as that of carbon or lead, the gluing strong force is so great that it doesn't matter that there's a lot of electrical power hidden away inside, trying to push the protons apart. It won't win. But in a big nucleus, a really huge one such as that of uranium . . . could the extra neutrons tip the balance?

Meitner and her nephew weren't physicists for nothing. They had paper with them, and pencils, and in the cold of the Swedish forest, this Christmas Day, they took them out and began calculating. What if it turned out that the uranium nucleus was so big, and so crammed with extra neutrons spacing apart the protons in there, that even before you start artificially pushing extra neutrons in, it's already in a pretty precarious state? It would be as if the uranium nucleus were a water droplet that already was stretched apart as far as it could go before bursting. Into that overstuffed nucleus, one more plump neutron was then inserted inside.

Meitner started to draw the wobbles. She drew as well as she played the piano. Frisch took a pencil from her, politely, and did the sketches. The single extra neutron that came in made the nucleus begin to stretch in the middle. It was like taking a water balloon, and squeezing it in the middle. The two ends bulge out. If you're lucky, the rubber of the balloon will hold, and the water won't burst out. But keep on with it. Squeeze in, and when the balloon spreads sideways, let go until it rebounds back toward the center and then squeeze in the opposite way. Keep on repeating. Eventually the balloon will burst. Get your timing right, and you don't even have to squeeze very hard. Each time the water balloon is rebounding back, you just let it reach its maximum rebound, and then—as with pumping on a swing—you give it a further squeeze to speed it on its way into yet another rubber-stretching contortion.

In the uranium nucleus, that's what the incoming neutrons had been doing. The reason Hahn had so much trouble classifying what he saw was that he'd been convinced adding neutrons would only make a substance heavier. Uranium atoms had been split for years, it seemed, and no one had realized it. But how to be sure?

There was one way to check it. The electricity of the protons within the nucleus could now be available to make the bits fly apart. In the units by which physicists keep count, that's about 200 MeV—200 million electron volts. Frisch and Meitner worked that out mostly in their heads. But would Einstein's 1905 equation prove that there really was that amount of energy available inside to send the nucleus roaring apart? Frisch takes up the story:

> Fortunately my aunt remembered how to compute the masses of nuclei ... and in that way she worked out that the two nuclei formed by the division of a uranium nucleus would be lighter than the original uranium nucleus, by about one-fifth the mass of a proton. Now whenever mass disappears, energy is created, according to Einstein's formula $E=mc^2$....

But how much energy would that be? One-fifth of a proton is a preposterously tiny speck of matter. The dot over a letter *I* has many more protons than there are stars in our galaxy. Yet that fifth of a proton—that subvisible speck—has to be enough to generate 200 MeV of energy. In Berkeley, California, a magnet was under construction, weighing as much as a large house—it was big enough to drive a truck under—which might, when charged with more electricity than the entire University of California ordinarily used, power up a particle to 25 MeV of energy. And now this speck was supposed to produce even more.

It would seem impossible—except for the immense size of c^2. The world of matter, and the world of energy, are linked by that massively widening bridge. From our perspective, the fragment of a proton slips across the roadway of that "=" sign: transforming; growing.

Growing.

They had crossed a river on their walk out from Kungalv, but it was frozen. The village was too far away to hear any market noises. Meitner did the calculation. Frisch remembered later: "One-fifth of a proton mass was just equivalent to 200 MeV. So here was the source for that energy; it all fitted!"

The atom was open. Everyone had been wrong before. The way in wasn't by blasting harder and harder fragments at it. One woman, and her nephew, quiet in the midday snow, had now seen that. You didn't even have to supply the power for a uranium atom to explode. Just get enough extra neutrons in there to start it off. Then it would start jiggling, more and more wildly, until the strong forces that held it together gave way, and the electricity inside made the fragments fly apart. This explosion powered itself.

Meitner and her nephew were still living in a world where science was politically neutral, so they prepared their discovery for publication. In naming it, Frisch was reminded of how bacteria divide. Back in Copenhagen he asked a visiting American biologist at Bohr's institute for the right word in English. The label *fission*, accordingly, to describe how atomic nuclei divide, was introduced in his ensuing paper. Hahn had already published the Berlin findings with minimal credit to Meitner, and soon began a nearly quarter-century-long campaign to pretend that all the credit was really his own.

$E=mc^2$ · $E=mc^2$ · $E=mc^2$

The thirty-year quest was over. In the decades since Einstein's equation first appeared in 1905, physicists had shown how the atom could be opened, and the compressed and frozen energy that $E=mc^2$ spoke about let out. They'd found the nucleus, and a particle called the neutron that could get in and out of the nucleus pretty easily (especially if one used the trick of sending it in slowly), and they'd found that when extra neutrons were pushed into overstuffed atoms such as uranium, the whole nucleus wobbled, and trembled, and then exploded.

What Meitner had realized was that this could occur because of the way the powerful electricity within the nucleus was held in by the springs or glue of the strong nuclear force. When an extra neutron made the nucleus start wobbling, those springs gave way, and the inner parts flew apart with wild energy. If you checked all the weights before and after, you'd find that the bits flying apart "weighed" less than when they were still held together in the nucleus. That "disappeared" mass was what powered their high-velocity escape. For it hadn't truly disappeared. The deep insights behind the equation and guaranteed that it would simply become apparent in the form of energy, getting the powerful c^2 boost to magnify it (in units of mph^2) by nearly 450,000,000,000,000,000 times.

It was an ominous finding, for in theory anyone could use it to start cracking apart nuclei, those central cores of atoms, and letting out these wild blasts of energy. In any other era, the next steps might have taken place slowly, with the first atomic bomb only appearing some time in the 1960s or 1970s. But in 1939, the world had just begun its largest war ever.

The race was on to see in which country the equation's power would emerge first.

Part Four . . .

. . . Adulthood

CHAPTER TEN

Germany's Turn

By 1939, Einstein was far from being the unknown young man whose father had to beg a Leipzig professor to give him a job. His work in relativity had made him the most famous scientist in the world. He had been Berlin's leading professor, and when anti-Jewish mobs and politicians had made it impossible to stay there, he went to America, in 1933, taking up a position at the new Institute for Advanced Studies in Princeton, New Jersey.

When Einstein heard of Meitner's results and how other research teams were beginning to extend it, colleagues were able to get one of the president's own confidants to carry a personal letter direct to the White House.

F. D. Roosevelt,
President of the United States,

White House
Washington, D.C.

Sir:

Some recent work ... which has been communicated to me in manuscript, leads me to expect that the element uranium may be turned into a new and important source of energy in the immediate future. Certain aspects of the situation which has arisen seem to call for watchfulness and, if necessary, quick action on the part of the Administration....

This new phenomenon would ... lead to the construction of bombs, and it is conceivable—though much less certain—that extremely powerful bombs of a new type may thus be constructed. A single bomb of this type, carried by boat and exploded in a port, might well destroy the whole part together with some of the surrounding territory....

Yours very truly,
Albert Einstein

Unfortunately, it met with this reply:

The White House
Washington

October 19, 1939

My dear Professor,

I want to thank you for your recent letter and the most interesting and important enclosure.
    I found this data of such import that I have convened a board.... Please accept my sincere thanks.

Very sincerely yours,
Franklin Roosevelt

Even someone who'd only been in America for a few years, as Einstein had, would understand that "most interesting" was a brush-off. Presidents are constantly sent impractical ideas. There's an obligation to be polite when the sender is famous, but FDR and his colleagues did not believe a bomb could possibly destroy a whole port.

The letter was shuttled away from FDR's desk, and ended up in the hands of Lyman S. Briggs, the easygoing, pipe-smoking director of the federal government's Bureau of Standards. He would be responsible for all U.S. atomic bomb development.

In the long history of governments assigning the wrong man to a job—and there have been some choice ones—this is one of the choicest. Briggs had entered government service during the administration of Grover Cleveland, in 1897, before the Spanish-American war. He was a man of the past, comfortable with that time when everything had seemed easier, and America had been safe. He wanted to keep it that way.

In April 1940, Meitner's nephew, Robert Frisch, then in England, began to convince the British authorities that a practical bomb could be built. A top-secret memo carrying the news was later rushed to Washington. By then there had been massive battles throughout Europe; panzer armies had overrun ever more countries. But you couldn't trick Lyman S. Briggs. That darn-foolish Brit report could be a danger if it ever got out. He locked it in his safe.

Germany's bureaucrats, even scientifically untrained ones, took the opposite view of history. What good was the recent past? It had only led to the sellout at the end of World War I, the corruption of the Weimar Republic, inflation, and then unemployment. The beckoning future would be better. That's why there was such belief in new roads, new cars; new machines—and new conquests. The latest laboratory speculations promised something new and powerful. Joseph Goebels later noted in his diary: "I received a report about the latest developments in German science. Research in the realm of atomic destruction has now proceeded to a point where . . . tremendous destruction, it is claimed, can be wrought with a minimum of effort. . . . It is essential that we be ahead of everybody. . . ."

And they had just the man for the job.

$E=mc^2$ $E=mc^2$ $E=mc^2$

In the summer of 1937, early in the month of July, Werner Heisenberg was on top of the world. He was the world's greatest living physicist after Einstein, famous for his work in quantum mechanics and the Uncertainty Principle. He had just been married, and now was returning after an extended honeymoon to the old family apartment in Hamburg, where his mother still lived, and the old five-foot-long electrically operated battleship he'd made as a teenager was still on display. He had a pleasant phone call to make, for he'd also been appointed to a senior position at the same university department where he'd earned his own Ph.D., as the wonder of the German academic establishment almost fifteen years earlier. He dialed the university's rector from his mother's phone.

Heisenberg had a way of standing with shoulders squared straight, in a state of alert excitement, whenever he was pleased. The call went through, but the rector told him there was a serious problem. An elderly physicist, Johannes Stark, had convinced the weekly magazine of the SS to run an anonymous article saying that Heisenberg wasn't sufficiently patriotic, that he had worked with the Jews, didn't have the proper pro-German spirit, etc.

This was the sort of public attack that often preceded a late-night arrest and then deportation to a concentration camp. Heisenberg was scared, but also furious. They were picking on the wrong man! It's true he'd worked with Jewish physicists, but Bohr and Einstein and the great physicist Wolfgang Pauli and so many others were Jewish that he'd had no choice. And despite that he'd always stood up for his country in public discussions, defending Hitler's actions; he'd always faithfully rejected the job offers from top foreign universities.

Heisenberg tried enlisting closest friends to help, but that had no effect. Soon he was brought for questioning to the basement of SS headquarters at Prinz-Albert-Strasse in Berlin, where the walls were uncovered cement, and the mocking sign "Breathe deeply and calmly" was up. (He wasn't beaten, and one of the interrogators had taken a Ph.D. at Leipzig for which Heisenberg was an examiner, but his wife later said he had nightmares about it for years.) Only when there were no signs of the SS attack letting up did he enlist one more ally, the woman who was closest to him of all: his mother.

The Heisenbergs were from the educated middle class, and so were the Himmlers, and Heisenberg's mother had known Heinrich Himmler's mother from the time they were young. In August, Mrs. Heisenberg went to see Mrs. Himmler, in her small but very clean apartment, where fresh flowers were always placed in front of the crucifix, and she passed along a letter from her son.

At first Mrs. Himmler didn't want to bother *her* son by delivering it, but as Heisenberg later recalled, his mother played the trump card: "Oh, you know, Mrs. Himmler, we mothers know nothing about politics—neither your son's nor mine. But we know that we have to care for our boys. That is why I have come to you.' And she understood that."

It worked.

From the office of the director of the SS

Very Esteemed Herr Professor Heisenberg!

Only today can I answer your letter of July 21, 1937, in which you direct yourself to me because of the article of Professor Stark....

Because you were recommended by my family I have had your case investigated with special care and precision.

I am glad that I can now inform you that I do not approve of the attack ... and that I have taken measures against any further attack against you.

I hope I shall see you in Berlin in the fall, in November or December, so that we may talk things over thoroughly man to man.

With friendly greetings,
Heil Hitler!
Yours,
H. Himmler

P.S. I consider it, however, best if in the future you make a distinction for your audience between the results of the scientific research and the personal and political attitude of the scientists involved.

The P.S., it was understood, meant that he could use Einstein's results on relativity and $E=mc^2$, but he had to show he disavowed Einstein himself, and gave no support to the sort of liberal or internationalist views—supporting the League of Nations; speaking out against racism—that Einstein and other Jewish physicists had been known for taking.

These terms weren't hard for Heisenberg to accept. In his teens he'd been active in the wandering clubs, where young Germans hiked for days or weeks at a time through wilderness areas, getting in contact with the true roots of their nation. Often, around the campfire, they'd talk of mystical heroes, and the way their country might regenerate in a hoped-for "Third" Reich, to be led—in the common phrasing of the time—by a sufficiently far-seeing leader, or *Fuehrer.* Many youngsters grew out of this, but well into his twenties Heisenberg remained drawn to the movement, despite mocking remarks from his more grown-up or liberal colleagues. During advanced university studies he'd regularly leave a seminar and meet up with a group of teenage boys, to lead them for a long walk and if possible a full night out in the woods, before racing back, by train, just in time for his next 9:00 A.M. lecture.

When the German army's Weapons Bureau began work, in September, 1939—a month after Einstein's letter to FDR—Heisenberg was one of the first to volunteer to do anything that was needed. The Reich was already at war: its artillery and ground troops and air force and panzers successful in Poland, but there were much greater enemies still to come. Heisenberg had always been an energetic worker, and now he surpassed himself. In December he delivered the first part of a comprehensive paper on how to construct a workable atomic bomb. In February, 1940, he had the full report done, and when a Berlin unit was set up to start building a reactor, in parallel with the work at his home university of Leipzig, he took command of both, regularly commuting between the two cities. It would have been tiring for most men, but Heisenberg was at the peak of his powers. He was still only in his thirties, a regular mountain-climber, a rugged horse rider, and an officer with two months' training each year in a crack alpine army unit.

The first tests were in Berlin, in an ordinary plank-walled building built on the wooded grounds not far from Meitner's old institute, under the cherry trees that bloomed so well in the warm, clear summer of 1940. To keep out the curious, the building was named "The

Virus House." Heisenberg's first step was to load in enough uranium—far more than the fractions of an ounce Meitner had arranged for Hahn to try in 1938. Only a very few atoms had exploded apart then. The sample had been so thin that most of the neutrons that the atoms released simply sped off into the lab's empty air.

Heisenberg ordered dozens of pounds of uranium. It wasn't hard to get, because the Reich's army had taken over Czechoslovakia a full year before invading Poland. Europe's largest uranium mines, which Marie Curie had once used, were at Joachimsthal there. The uranium was delivered. With the prestige of Heisenberg's name, the Weapons Bureau could ensure there were priority train shipments.

Simply stacking together a large amount of uranium wouldn't get a reaction going. For a nucleus, as we saw, is a vanishingly tiny speck, deep in the empty spaces inside an atom. Most of the neutrons released by the first explosions would speed right past the nucleus, like an alien space probe hurtling through our solar system.

The twist Fermi had found—that there was a great power in using slowed neutrons—could help solve this problem, and get a reaction to begin. Fast neutrons can be thought of a sleek, "streamlined" as they travel. But as we saw, slow neutrons can be thought of as more "wobbly"; more spread out. Even if their main body only comes close to one of the waiting nuclei, then part of the neutron—the spread-out bit—is likely to connect. What would have been a near miss of a nucleus if the neutron were coming in fast, becomes far more likely to be a "catch." When such a slow neutron gets caught or pulled in, circumstances are prepared for $E=mc^2$ suddenly to "operate": for the nucleus to wobble, then explode, letting out its great blasts of energy, and also, in the process, letting out a gush of extra neutrons, so that yet more atoms are hit, to wobble and explode apart in turn.

Heisenberg searched for the right substance to produce that useful slowing of the neutrons. Anything that was dense with hydrogen would work to some extent, because the "bumpings" of neutrons against hydrogen tends to slow the neutrons down. That was why Fermi, back in 1934, had managed to get some effect even with ordinary water (H_2O) from the goldfish pond on the grounds of his institute. But when the first German teams tried spreading ordinary water around a uranium sample, although there were a few crackling reactions at the center of the sample, where the first uranium atoms broke apart, the neutrons that gushed out were still moving too fast to get the reaction to spread.

Heisenberg needed a better moderator. He knew that around the time when Fermi had been working, a U.S. chemist, Harold C. Urey, had discovered that the water in all the world's oceans and lakes is not composed merely of H_2O. Mixed in with it is a variant molecule, slightly heavier. Instead of having ordinary hydrogen at the core, it has deuterium, which is very similar to hydrogen but weighs about twice as much. Aside from that it is the same as ordinary water—it's just as free-flowing, and transparent; it's part of our rain and ice and seas; we drink it all the time. But there's only one molecule of it for every 10,000 molecules of ordinary water, which is why "heavy water" had been overlooked before. (A large swimming pool has only about a drinking glass's worth spread within it.) But heavy water is superb at slowing down high-velocity neutrons. They smack into the heavier deuterium and start ricocheting, slowing with each bounce, and emerge a fraction of a second later, after several dozen ricochets, moving much slower than when they went in.

German labs had accumulated only a few gallons of heavy water at first. This wasn't enough to be shared between Leipzig and

Berlin. Heisenberg's sentiments were more with Leipzig, so it was in the basement of the physics institute that the most important tests were prepared. As 1940 went on, the precious heavy water was poured in with the several pounds of stacked uranium. The mix of heavy water and uranium was packed into a tough spherical containment vessel, then dangled from a hoist while measuring devices were set up around it. Professors weren't supposed to get involved with the minutiae of experiments, but Heisenberg prided himself on his practical skills as much as on his theoretical gifts. He had built some of those measuring devices himself, with Robert Dopel, the main experimenter in charge.

When the uranium and heavy water were in place, and the measuring devices set up closely around the container, the experiment was ready to begin. Gunpowder needs a match to start. Dynamite needs a blasting cap. For an atomic reaction, even one that's in uranium of too low a quality to lead to a full explosion, there has to be an initial source of emerging neutrons. Dopel had left a hole in the bottom of the containment vessel. The neutron source was a small amount of radioactive powder, similar to the one Chadwick had used. It was brought over, contained in a single long probe, and now, for the first time—in an ex-

periment that began in February 1941—all
the working parts that could come together in
a bomb were in place.

When Dopel and Heisenberg gave the instructions, they expected the probe to go in,
and the first speeding neutrons would be let
loose inside the uranium. A few of the uranium atoms would burst, flying apart much
faster than anyone would have suspected before Meitner explained how $E=mc^2$ was operating. Even more neutrons would spray out in
the fast debris. They'd pass without much effect through the first layers of uranium atoms
they hit, but when they reached the heavy water they'd be caught bouncing back and forth
till they slowed down. When they went on to
the next layer of uranium metal they'd be
wobbling forward so slowly, and so dis-
persedly, that they would be much more likely
now to connect with the uranium nuclei, and
especially the most fragile ones, and overload
them, making them wobble, and tremble, and
then explode apart in turn.

Each one would be another crackling oc-
currence of $E=mc^2$, in a sequence that Geiger
counters would show building up faster and
faster. In the first few millionths of a second—
by Heisenberg's calculations—there would be
perhaps 2,000 explosions. In the next millions
of a second there would be 4,000. Then

8,000, then 16,000, and so on. Doubling on that time scale rushes upward very quickly. If everything worked, there would be trillions of these minute explosions still in only a fraction of a second, and then hundreds of trillions, and the cascading effect would keep on going. It would be a "rip" in the ordinary fabric of matter, as all the energy that had been squeezed inside these atoms for billions of years now came out: there in the basement of the Leipzig Institute; in this university run by officers appointed from Reich headquarters; with students proudly wearing the swastika in the classrooms above. To explode apart a billion atoms, you wouldn't need to set up an enormous laboratory, with a billion initiating neutron machines. Once a very few atoms started, the debris fragments they sent out—loaded with neutrons—would quickly start up the rest. The uranium wasn't purified enough to produce a runaway reaction, but it would be a start.

The professors gave the instruction. Dopel's assistant, Wilhelm Paschen, inserted the probe. It was early 1941. The initiating neutrons were inside the uranium! Everyone stared at the dials to record the result.

And nothing happened.

There wasn't enough uranium to get a reaction going. Heisenberg wasn't fazed, and simply ordered more, from the enterprising Berlin Auer company, which over the years had moved on from toothpaste, and was a wholesale supplier of a whole range of uranium products. Its raw supplies were no problem, as Einstein had also warned in his letter to FDR "Germany has actually stopped the sale of uranium from the Czechoslovakia mines which she has taken over," Einstein had written, ". . . while the most important source of uranium is the Belgian Congo."). The Union Miniere in occupied Belgium had thousands of pounds stockpiled from those Congo mines. When the Joachimsthal stockpiles ran low, the Germans turned there next.

Machining uranium into a usable form was hard, since it demands a lot of labor, plus the fine uranium dust that's produced is dangerous for the workers. But Heisenberg had a procurement organization to draw on that wasn't hindered by outmoded notions of human rights. Germany had many concentration camps, full of people who were soon to be killed anyway. Why shouldn't important projects get some advantage from them? As the war went on, Berlin Auer executives calmly bought female "slaves" from the Sachsenhausen concentration camp. They could be worked to prepare the uranium oxide the German project needed. Back in April

1940, Heisenberg had expressed his impatience at how long the first Auer shipments were taking. The administrative head of the army project assured him that Berlin Auer would have its workforce operating at maximum speed. Supplies had started coming in that summer, and now in 1941, even more was quickly shipped.

In autumn 1941 there were more promising tests, and finally, in spring 1942, the breakthrough happened. The containment vessel was pouring out neutrons: 13 percent more than the inserted source had pumped inside it to start with. The trapped energy that Einstein had first spoken of nearly 40 years before was now being released. It was as if a narrow funnel was stretching up from deep underground, and a thundering wind—the released energy—was blowing fast along it. Himmler's faith was justified.

Heisenberg—triumphant—had managed to get the power of Einstein's equation to come roaring up, and appear in Nazi Germany.

$$E=mc^2 \cdot E=mc^2 \cdot E=mc^2$$

Einstein was getting hints of Heisenberg's success, for Meitner's successor as director of the Kaiser Wilhelm Institute for physics had

been Dutch, and when he in turn was expelled, ending up in America, he shared with
his new colleagues what he'd heard of he work
at The Virus House and Leipzig.

Einstein wrote another letter to FDR: "I
have now learned that research in Germany is
carried out in great secrecy and that it has been
extended to another of the Kaiser Wilhelm Institutes, the Institute of Physics." But this time
it seems that there wasn't even the courtesy of
a reply. A white-haired foreigner is one thing,
especially if he has a grand reputation in science. But tensions were rising as war got
closer, and the FBI now saw reasons to dis-
count anything he said. For Einstein was a so-
cialist, and a Zionist—and he had even spoken
out against excess profits for arms manufactur-
ers. The FBI reported back to army intelligence that:

> "In view of his radical background, this of-
> fice would not recommend the employ-
> ment of Dr. Einstein, on matters of a secret
> nature, without a very careful investiga-
> tion, as it seems unlikely that a man of his
> background could, in such a short time,
> become a loyal American citizen."

When the United States finally did get a serious atomic project started, it was helped

through some skillful manipulations by impatient visitors from Britain. Mark Oliphant was another one of Rutherford's bright young men, and in the summer of 1941 he led a two-front assault. First he arrived in Washington, dangling the gift of the cavity magnetron—a key device for shrinking room-sized radar sets to a volume that could be crammed into an airplane. (This was when Oliphant discovered that Lyman S. Briggs, leader of the West's atomic research project, had locked the top-secret British results inside his safe.) Next, Oliphant traveled to Berkeley, where the physicist Ernest Lawrence worked.

Lawrence was not especially bright as physicists go, but he loved machines, great big powerful machines, and his very simplicity—his directness of focus—allowed him to get them built. For example, Samuel Allison (working at the University of Chicago then) remembers that Briggs had "a tiny cube of uranium which he liked to keep on his desk and show to insiders. . . . Briggs used to say, `I want a whole pound of this,' . . . Lawrence would have said he wanted forty tons and got it."

By the autumn of 1941 Briggs was out, and a group of more effective leaders including Lawrence was in, and by December—when Pearl Harbor brought the United States into the war—the project really took off. It came to be called the Manhattan Project, as part of the cover story that it was simply part of the Manhattan Engineering District.

The refugees Briggs had scorned were indispensable. Eugene Wigner, for example, was a remarkably quiet, unassuming young Hungarian, who came from an equally quiet and unassuming family. When World War I had broken out, Eugene's father had stayed away from political discussions, pointing out, quite sensibly, that he was pretty sure the emperor was not going to be swayed by the views of the Wigner family. But this caution meant that when Eugene, a superb student, was facing university choices, the father had him take a practical engineering degree, as the odds on a career in theoretical physics taking off were very slight.

Wigner did succeed at physics, and after he was forced out of Europe in the 1930s, he ended up centrally involved in the American duplicate of Heisenberg's calculations, detailing how a reaction could begin. But his engineering training meant that he handled the subsequent steps far better than Heisenberg. What shape, for example, should the uranium be that would go inside a reactor? The most efficient possible design would be a sphere. That way the maximum number of neutrons would be deep in the center. Next best—if a sphere was too hard to cut accurately—would be an oval shape. After that comes a cylinder, then a cube, and last, worst of all, would be to try building it with the uranium stretched out in flat sheets.

For his Leipzig device, Heisenberg had chosen the flat sheets. The reason was simply that flat surfaces almost always have the easiest properties to compute, if you're advancing by pure theory. But engineers with enough practical experience are never restricted to pure theory. There are many informal tricks of the trade for how ovals and other shapes work. Wigner knew them, as did many other similarly cautious refugees, who'd also been advised by their families to take engineering degrees. Heisenberg did not. That was of central importance. Professors in general tend to be hierarchical, and pre-World War II German professors were at the peak of such confidence. As the war went on, a number of junior researchers in Germany found that Heisenberg had been mistaken in one engineering assumption or another. But Heisenberg almost always refused to listen; would angrily try to keep them from even daring to mention it.

Even so, nobody could be confident the United States was going to win the race to make the bomb. America was just coming out of the Great Depression; much of its industrial base was still rusted and abandoned. When Heisenberg began his research for army ordnance, the Wehrmacht was the world's most powerful fighting force. It had entire army groups supplied with equipment that surpassed that of any other nation. The United States had an army that, if you included a lot of generation-old World War I artifacts, could just about supply two divisions, thus ranking it seventeenth in the world, a little behind Belgium.

Germany also had the world's best engineers, and a strong university system—despite having expelled so many Jews—and above all, they had that head start: two precious years when Heisenberg and his colleagues had been working full out, while Briggs had mostly been musing at his desk. These were the quirks of fate that would influence who ended up using the equation first. $E=mc^2$ was far from the pure reaches of Einstein's inked symbols now. The Allied effort would have to go faster.

The German effort would have to be sabotaged.

CHAPTER ELEVEN

Norway

British intelligence had been monitoring the German program from the beginning, and identified its one weak spot. It was not the uranium—there was too much in Belgium to try to destroy, even if anyone could get at it. Nor was it Heisenberg himself: no assassination squad could reach him in Berlin or Leipzig, and even his summer resort in the Tyrols was too far away, and probably too well guarded as well.

The most vulnerable target was the heavy water. A reactor couldn't fully ignite without slowing down the neutrons from the first atom explosions, making them "spread out" so they could find the other nucleus specks, and jostle them to start exploding their hidden energy in turn. Heisenberg had decided on heavy water for that, but it takes a very large factory—using a great amount of water—to separate that from ordinary water.

Some cautious members of Heisenberg's staff had proposed that Germany construct a factory of its own, safely on German soil, to produce the heavy water. But Heisenberg, backed by army officials, knew there already was a perfectly sound heavy water factory in operation, using the abundant, power-generating waterfalls of Norway. It's true that until recently Norway had been an independent country, but now wasn't it merely a conquered province?

It was a fateful decision, but generations of German nationalists had felt their country was suffocated, entrapped. Heisenberg backed the decision to rely on the Norwegian factory, for he backed the idea of the new Reich's right to dominate all of Europe. Through the war he excitedly visited one subject nation after another; striding through the offices of his one-time colleagues, local collaborators accompanying him; in the Netherlands explaining to the aghast Hendrik Casimir that although he knew about the concentration camps, "democracy can't develop sufficient energy," and he "wanted Germany to rule."

The Norwegian factory was located up in a mountainous ravine, at Vemork, 90 miles by winding road from Oslo. Before the war it

had produced only 3 gallons—24 pounds—
of heavy water a month, for laboratory research. Engineers from the great I. G. Farben
industrial combine in Germany had asked for
more, and offered to pay above market rates,
but the Norwegian managers had refused,
unwilling to help Nazis. A few months later
the Farben engineers asked again; this time—
for the Wehrmacht had destroyed the Norwegian army—they were backed by troops
with machine guns. The Vemork staff had no
choice but to agree. Production had been accelerated to an annual rate of 3,000 pounds
by mid-1941. Now, in mid-1942, it was up
to 10,000 pounds per year, steadily shipped
to Leipzig, Berlin, and the other centers for
atomic research.

There were only a few hundred troops
guarding the factory, for the site seemed impregnable. The Norwegian resistance was
clearly too small and untrained to be feared for
an assault on such a huge factory. The complex
was surrounded by barbed wire and arc lights,
with only a single suspension bridge giving access. It was located in a setting so deeply cut in
the mountains that for over five months of the
year shadows from the surrounding peaks kept
direct sunlight entirely away, and workers had
to be taken up by cable car, to a higher plateau,
to get a daily dose of light.

This was the target the British government chose to attack. If Vemork had been on the coast, then members of the Royal Marines could have tried to go in, but since it was 100 miles inland, a team from the First Airborne Division was chosen. These troops were good. Many were working-class London boys, fists trained from surviving the Depression, and now in their twenties, undergoing more serious training: weapons, radios, explosives. They weren't told where they were going, of course—that would only come on the day of the mission. Until then they believed they were being training for a paratrooper competition with the Yanks. That their fate was being directed in an effort to control what Einstein's equation and Rutherford's investigations was leading to—of that they had no idea at all.

Two glider teams took off after dark from northern Scotland, towed by the new high-speed Halifax bombers. There were about thirty troops in all. (Today we think of a typical glider as a single-man device, but then, before helicopters were widespread, they were often much bigger, resembling small cargo airplanes without motors.) It was a terrible night. Huge ore deposits in the mountains they passed seem to have disoriented the compass of one of the planes, guiding the pilot into a mountain edge.

The pilot of the other team's glider was an Australian, who found himself caught in an impossible dilemma in this disorienting northern hemisphere snowstorm at night: if he stayed with the towing Halifax when it was high, his wings and cable lines iced up so heavily that he would crash. But if he released and flew low too early, swirling gales in the mountains would toss him away from any controlled path. The Australian's glider finally released, in heavy cloud, but something went wrong and it too came down in a heavy crash landing.

At each crash site there were a number of survivors, and in both cases a few of the troops—injecting themselves with morphine for their injuries; popping amphetamines to get through the snow—managed to reach local farmhouses for help. But all were soon arrested by German troops or local collaborators. Most were shot immediately; the others were tormented for a few weeks first.

$E=mc^2$ · $E=mc^2$ · $E=mc^2$

Just a few years earlier, R. V. Jones had been a promising astronomy researcher at Balliol College, Oxford. Now, barely out of his twenties, he was director of intelligence on the air staff: faced with the sort of ethical dilemma that is an occasion for cleverness at Oxford College; haunting in real life. Thirty Airborne specialists had been sent in, and every single one was dead. The factory hadn't even been reached.

"It fell to me," Jones remembered decades later, "to say whether or not a second raid should be called for. It came all the harder because I should be safe in London, whatever happened to the second raid, and this seemed a singularly unfitting qualification for sending another 30 men to their deaths. . . .

"I reasoned that we had already decided, before the tragedy of the first raid and therefore free from sentiment, that the heavy water plant must be destroyed; casualties must be expected in war, and so if we were right in asking for the first raid we were probably right in asking that it be repeated."

This time the Norwegians themselves took over. Six volunteers who'd made it to Britain were selected. One was an Oslo plumber, another had been an ordinary mechanic. Contemporary records suggest that the British had little confidence the Norwegians would succeed where dozens of crack Airborne troops had failed. Minimal attention, for example, was given to their possible escape afterward. But what else was there to do? As more heavy water continued to be shipped to Germany, work in Leipzig could progress; the Virus House unit in Berlin could catch up as well.

The six Norwegians were trained as well as possible, then sent to a luxurious safe house—S.O.E. Special Training School Number 61—outside Cambridge for final preparations, but mostly to wait for the weather to clear. There were chatty English girlfriends, and the occasional dinner out in Cambridge. Then in February 1943 the meteorological reports improved; the house suddenly emptied.

After being parachuted in to Norway, they met up with an advance party of a few other Norwegians, which had waited in isolated huts all winter. Together, on cross-country skis, they reached Vemork a few weeks later, at about 9 P.M. on a Sunday night.

"Halfway down we sighted our objective for the first time, below us on the other side. . . . The colossus lay like a medieval castle, built in the most inaccessible place, protected by precipices and rivers."

It was the furthermost ripple of what had begun in Einstein's quiet thoughts: a handful of armed Norwegian men, panting in deep snow, staring at a lit fortress in the night. It was clear why the Germans had left only a small guard. The only way in was across the single suspension bridge, over an impassable stone gorge several hundred feet deep. It might be their protected emplacements, but if that happened, the Germans would simply start killing the local townspeople. Both sides knew this. When a radio transmitter had been uncovered on Telavaag island the year before, every house and boat was burned, and all the women, and all the children—and of course all the men—who'd lived on the island were sent to concentration camps. Jones in London would probably not accept that again; the nine Norwegians looking down on the factory now definitely wouldn't. But this didn't meant they were going to go back. They had another way in.

From aerial reconnaissance photographs, highly magnified in England, one of the team—Knut Haukelid—had noticed a clump of scrub plants a little further along the gorge. "Where trees grow," he'd remarked, "a man can make his way." One of their members had reconnoitered the day before, to confirm this. They started the climb down, cursing their heavy backpacks, then crossed the river, which was ominously oozing water above its ice, and then cursed their backpacks even more on the climb up to the factory. Since no one wanted to disappoint the others, they all furtively quickened their pace, so that the speed was soon exhausting.

Outside the factory perimeter they had to rest, sharing chocolate to get some strength. There was a loud noise of turbines, for due to the orders from Leipzig and Berlin, the factory worked on a twenty-four-hour schedule. What do nine highly armed men talk about? One was teased for how he was trying to pick rations out from between his teeth without the others noticing; others spoke, more seriously now, about two young married couples they'd met on the final night before their skiing journey to Vemork. One of the parachuted fighters had been at school with the young man in one of the couples, but at first they'd been scared at coming across armed strangers: they hadn't recognized him. Then when they finally did, each side had realized it was too dangerous to talk, even though the parachuted newcomers were desperate to hear what ordinary life had been like in Norway this past year. They'd had to spend the night aware of the lamps on in the couple's cabin, and the sight of smoke from their hearth fire; busying themselves so they would have no thoughts of home; just checking rifles and grenades and explosives, and waxing their cross-country skis for this assault.

One of the men looked at his watch; the short rest was over. They lifted their packs, and went to the gates. There were advantages to having a big ex-plumber with them, for he now took out oversized wire-cutters and snapped right through the iron. They were inside.

It was the central moment. Heisenberg and the German army's weapons bureau had been constructing a "machine": a vast apparatus composed of uranium, and trained physicists, and engineers, and electricity supplies, and containment vessels, and neutron sources. Only when every part was in place could the mass from the center of uranium atoms be sucked out of existence, to be replaced by roaring energy in fast, unstoppable $E=mc^2$ explosions. The heavy water that controlled the flight of the triggering neutrons, slowing them down enough to "ignite" the uranium fuel, was the last part of the machine that had to be put in place. Germany's power—of troops and radar stations and local collaborators and SS inquisitors—had swatted down the British Airborne forces that had tried to obstruct the "machine" that would allow the power of $E=mc^2$ to emerge.

The nine Norwegian men were now all that were left. One group took up positions outside the guard barracks. Others watched the huge main doors to the factory. Blasting those open would have been possible, but again would have resulted in reprisals. An engineer who'd worked at the factory, though, had told the Resistance about a little-used cable duct that went in from the side. Two of the team, now loaded with all the explosives they'd carried, found it and crawled in.

The workers inside had no love for I. G. Farben, and were only too willing to let them go ahead. Within about ten minutes the charges were set. The workers were sent out, and the two men quickly followed.

At about 1 A.M., there was a slight thud; a brief flash at a few of the windows. The eighteen "cells" that separated out the heavy water were thick steel and chest high, looking a bit like an overbuilt gas-fired boiler. No explosives that nine men could carry in their climb would totally destroy them. Instead, the Norwegians had set small plastic charges at the bottom of each one. The charges opened up holes, and also sent enough shrapnel flying out to cut exposed pipes.

The warm wind known as the *foehn* had started blowing, and the Norwegians could feel the snow starting to melt on their way back down the gorge. Searchlights came on as well as the air raid sirens, but this didn't matter. The terrain was rough enough to cover the men. As they climbed and then skied away, the heavy water gushed from the factory's drains, rejoining the mountain's streams.

CHAPTER TWELVE

America's Turn

The raid bought time for the Allies, but even that would have been wasted if the wrong person had headed their project to build a bomb. At one point the Berkeley physicist Ernest Lawrence's name had been mooted, but his personnel skills made Heisenberg look considerate. America's own physics establishment had been so weak in the 1920s and 1930s that any bomb would have to be constructed, in large parts, by more highly skilled refugees from Europe. No one could have been worse to lead such a team than the broad-shouldered South Dakotan Lawrence.

In 1938, the Italian refugee Emilio Segrè had obtained a position at $300 per month in Lawrence's lab. It was a godsend for Segrè, who was Jewish, since if he and his young wife had to go back to Italy, there would be no possibility of working in a university anymore; there was also a good chance that they would be turned over to the Germans, and—as happened to many of their relatives—their children would quite possibly be killed. Segrè recalls how Lawrence responded:

> In July 1939, Lawrence, who by then must have realized my situation, asked me if I could return to Palermo. I answered by telling him the truth, and he immediately interjected: "But then why should I pay you $300 per month? From now on I will give you $116."
>
> I was stunned, and even now, so many years afterward, I marvel … that he did not think for a second of the impression he conveyed.

The man who was appointed to the overall charge of the atomic bomb program, Leslie Groves, was somewhat better than Lawrence, at least in the sense that he wasn't prone to threatening his staff with imminent death. He also—like Lawrence—was effective in getting things built. He'd finished fourth in his class at West Point and after a stint at MIT, had been largely responsible for getting the Pentagon building completed. Before the atomic bomb project was done, a vast reactor would need to be built, sited by a large river to take away the cooling water; factories thousands of feet long would have to be constructed, able to filter toxic uranium clouds. Groves got them all done, on time and under budget.

But Groves also carried a constant personal anger, of a sort more accepted in American public life at the time. He screamed, he threatened; he demeaned his assistants in public; his neck veins popped out with anger a lot. (The fact that he was now dealing with theoretical physicists of an intellectual level that dwarfed the accomplishments he'd been so proud of at West Point did not make him any easier to live with.)

When the secret Los Alamos, New Mexico, research center for the bomb project officially opened, in April 1943, Groves stood up to speak. A young member of the audience, Robert Wilson, later remembered: "He said he appeared not to believe in the eventual success of the project. He emphasized that if—or when—we failed, it could be he who would have to stand before a congressional committee to explain how money had been squandered. He could not have done worse at starting the conference on an upbeat note of enthusiasm."

Many possible projects have failed when administrators like this took over. A workable jet engine, for example, had been up and running in Britain as early as 1938, but incompetent organization kept it from ever being deployed in sufficient numbers to help the RAF. Groves could motivate construction engineers who had to follow blueprints, yet would almost certainly have failed at inspiring theoreticians who had to trust that they would succeed in unexplored intellectual terrain. But in the autumn of 1942—while Heisenberg was readying further work after his successful Leipzig tests—Groves made an appointment of genius. He selected the exquisitely oversensitive J. Robert Oppenheimer to be in day-to-day control of the scientists at Los Alamos.

It was a job that nearly destroyed Oppenheimer's health; by the time the first explosion took place, Oppenheimer, about six feet one, was down to about 116 pounds. In time his work on the Manhattan Project destroyed his career, making him so much of an outcast in the U.S. that he would have been jailed if he tried to read his own past classified papers. But he got the job done.

Oppenheimer's great strength, curiously enough, came from his underlying lack of confidence. It wasn't something most people could tell on the surface, of course. He had graduated from Harvard in only three years, with perfect grades; had studied at Rutherford's lab; taken his doctorate from Gottingen, and quickly after, still in his twenties, become one of America's top theoretical physicists. He seemed effortlessly good at everything. He once asked a graduate student, Leo Nedelsky, to take over some of his lectures at Berkeley: "It won't be any trouble," said Oppenheimer, . . . "it's all in a book." Finding that the book was in Dutch, which he could not read, Nedelsky demurred. "But it's such easy Dutch," said Oppenheimer.

But it was all a fragile, frantic, uncertain ability. His whole family had been like that. His father had climbed up in the New York rag trade, then married a genteel woman who insisted her family do everything "properly": there were summerhouses, and servants, and classical music. At summer camp she saw to it that the other little boys were instructed to play with her Robert, and was surprised that he ended up being bullied, on one occasion being locked naked in the icehouse overnight. At Rutherford's lab he'd been so desperate at not being the top researcher that in a fit he'd tried to strangle his one friend. At Gottingen he'd had books hand-bound for himself, chided a graduate student couple for what he called their "peasant" ways in not being able to afford a baby-sitter—and then agonized over why people thought he was putting on airs.

As a result, Oppenheimer was superb at identifying weaknesses of inner doubts in others. When he lashed out at fellow researchers throughout his time as a Berkeley professor, he could unerringly select the one single area they felt weakest about, for he knew very well what it was to have an area to feel weak about. Even in his own physics he knew his own weaknesses, and felt a crushing sense of self-loathing at the way he regularly pulled back, ever so slightly, just when he might make a major breakthrough.

And then, at Los Alamos, he switched. The sarcasm was dropped, for the duration of the war. But the ability to detect other people's deepest fears or desires remained—and this meant that he became a superb leader of men.

He knew—instantly—that the young postgraduate physicists he needed in large numbers wouldn't pass up work at MIT's radar lab or other famous wartime projects to head to this unknown New Mexico site, simply on the basis of salaries, or offers of future jobs. They'd come only if they thought the top physicists in America were going there. Oppenheimer, accordingly, recruited the senior physicists first; the postgrads followed fast. He even got the authority-resistant genius Richard Feynman on his side. (Tell Feynman that something was a national emergency and his country needed him, and he'd give his mocking Brooklyn snort and tell you to get lost.) Oppenheimer understood that Feynman was so hostile in large part out of furious anger: his young wife had tuberculosis, and in the era before antibiotics it was likely she would soon die. But Oppenheimer obtained a rare-as-gold wartime train pass so she could come to New Mexico; he also arranged a place in a hospital close enough to Los Alamos so that Feynman could visit her regularly. In his later memoirs, Feynman joyously mocked every administrator he worked for— with the exception of the two years at Los Alamos, where he did everything Oppenheimer asked.

Oppenheimer's skills came to the fore in the hardest problem Los Alamos needed to solve. America was building two entirely different sorts of bombs. One team, led by Lawrence in Tennessee, took a blunt approach, and was simply trying to extract the most explosive component in natural uranium. When enough of that was accumulated, there'd be a bomb. The Tennessee factories followed the sort of straightforward engineering that Lawrence and other plain-talking Americans liked. Although there were exceptions, it was largely pushed by native-born Americans.

Another team, in Washington State, was taking a more subtle approach. They were staring with ordinary uranium, and then hoping to transform that, in a process of transmutation much like the one medieval alchemists and even Newton had struggled with in past centuries. The alchemists had wanted to turn lead into gold. The Washington State team, if they succeeded, would transform ordinary uranium into the wickedly powerful plutonium metal. Although again there were exceptions, this abstruse approach had been promoted more by the European refugees, educated in a more theoretical tradition.

The Pentagon liked Lawrence and the blunt Americans down in Tennessee, but it turned out that the foreigners' Washington project did best of all. Despite all of Lawrence's screaming and haranguing and threats, by early 1944 the Tennessee factories—giant factories, over a mile in total length; costing over a billion dollars (even in 1940s currency)—could barely sieve out enough purified uranium to stuff into a single envelope. No one was going to be able to make a bomb with that.

But although the Washington team did manage to create its promised plutonium, pretty soon the Los Alamos staff realized that no one could get it to ignite as a bomb. The problem wasn't that plutonium didn't explode. Rather, it exploded too *easily*. To make a simple uranium bomb—if the Tennessee team ever got enough purified uranium together—wouldn't be hard. If the amount that would make an explosion was 50 pounds, then you could make a 40-pound ball, and carve a hole in it, and then get a big gun, aim it at the hole, and fire—fast!—the remaining 10 pounds into it. The threshold would then be reached so quickly, and the reaction would take place in such a small concentrated area, that a very large amount of the explosive U235 form of uranium would convert into energy before it blew itself apart.

Plutonium was different. Fire two segments together, and the plutonium would start exploding before the two halves completely clanged together. You wouldn't want to stand nearby when this started, of course, for there would be a gush of liquefied or gaseous plutonium where the reaction began. But that would be all. There would be almost no nuclear reaction: most of the raw plutonium, not transformed, would simply spatter away.

This is where Oppenheimer's insight and managerial gift came in. Forget about trying to clang two separate pieces of plutonium together. The way to get the plutonium fuel from Washington State to work, he realized, would be to start with a ball of plutonium that was fairly low density. That wouldn't explode. But then you'd wrap explosives around it, and set them off, all at *precisely* the same instant. Do it right, and the ball would crumple inward, so fast that the trillions upon trillions of $E=mc^2$ blasts that started spreading within would have enough time to accumulate before the plutonium flew apart.

The technique was called implosion, but the calculations were so hard—how do you make sure the plutonium ball doesn't crumple unevenly?—that there was a great deal of cynicism about whether it could work. (When Feynman first saw what the implosion theorists were trying, he pronounced simply: "It stinks!") Oppenheimer overcame that. He nurtured the first theorists who proposed implosion; he assembled the right explosives experts; as the project grew to a level that under anyone else's supervision it might have fallen apart in a mess of squabbling egos, he deftly manipulated the participants so that all the different groups involved worked together in parallel.

At one point he had the top U.S. explosives expert and the top UK explosives expert and the Hungarian John von Neumann—the quickest mathematician anyone had met, who would also help create the computer in his long career—and a host of other nationalities all working on it. He even had Feynman joining in! The one prima donna who might have destroyed the effort was the embarrassingly egocentric Hungarian physicist Edward Teller. Oppenheimer neatly led him away, and granted him his own office and work team, even amid the shortages of skilled manpower, to concentrate on his own prize ideas. Teller was vain enough—as Oppenheimer of course understood—that he simply took it as his due; in his pleasure he no longer bothered everyone else.

Paralleling the whole team was a purely British effort, which touched on these theoretical matters as well as practical isotope separation, at Chalk River, near Ottawa. Groves had been suspicious of this group, but Oppenheimer wanted all the help he could get.

Money didn't count. Everyone knew the level Germany was starting from. At one time, at Los Alamos, calculations suggested that a casing of solid gold might help bounce escaped neutrons back into an exploding bomb. (Its weight would also help keep the exploding plutonium bomb intact.) A little later, Charlotte Serber, who ran the library cum document storage room at Los Alamos, received a small package; about the size of a brown paper lunch bag.

All that day Serber amused herself and the women who worked for her by saying to innocent would-be readers 'Please move these little packages to the next table for me.'

They couldn't. The package came from Fort Knox. Gold is denser than lead (that's why it was chosen), and the 6-inch solid gold sphere inside weighed eighty pounds.

But yet, despite the dozens of top researchers and the nearly unlimited funds, the plutonium problem still wasn't being solved. Was it possible, Oppenheimer and others worried, that no full bomb could be made this way? In that case, the best that might result would be an accumulation of radioactive plutonium. Was this what Heisenberg's heavy water reactor would cook up? Oppenheimer was informed, in a memorandum of August 21, 1943:

> ...It is possible ...that the Germans will have a production, let us say, of two gadgets a month. This would place particularly Britain in an extremely serious position but there would be hope for counter-action from our side before the war is lost....

One of the memo authors was Teller, who could be discounted, but the other was Hans

Bethe, an eminently sensible man. He was the head of the theoretical division at Los Alamos, and he'd been a student of Geiger's until 1933 in Berlin. He had excellent contacts with physicists who'd remained on the Continent. The "gadgets" Bethe and Teller had in mind were full bombs, which were unlikely at this stage, but who knew what else the Germans might build?

Even a few pounds of powdered radioactive metal released over London could make parts of that city uninhabitable for years. There already were worrying reports of the advanced delivery weapons Germany was working on, and one of Heisenberg's men was later seen at Peenemunde, where the supersonic "vengeance" weapon—the V-2 missile—was being built. Much simpler jet drones were also being constructed—the V-1s—and if those crashed highly radioactive warheads into Allied troop emplacements, in the south of England before D-Day or in France afterward, there could be casualties of a level that had never before been seen.

The threats were taken so seriously that Eisenhower accepted Geiger counters, and specialists trained in their use, to be ready to go with his troops building up in England for D Day. And then, at the very end of 1943, when Oppenheimer was most lost in the plutonium implosion problem, Niels Bohr arrived at Los Alamos, after an escape from his institute in Copenhagen. Bohr was the kindly elder statesman of physics. Over the years everyone who counted, from Heisenberg to Oppenheimer to Meitner's nephew Robert Frisch, had stayed at his institute and worked with him.

Now Bohr brought serious news. On December 6—after he'd fled—German military police had invaded his institute. They hadn't managed to steal the Nobel gold medals stored there, for George de Hevesy had dissolved them in a jar of strong acid, and left them—in liquid suspension—unobtrusively on a back shelf. But they had bullied their way around, arresting one of Bohr's colleagues who lived in the building. Most seriously, there were rumors that the institute's powerful cyclotron, an early form of particle accelerator, was going to be broken apart and sent back to Germany. Cyclotrons can make plutonium.

And then British military intelligence reported that the factory at Vemork had been restarted. I. G. Farben engineers had been working frantically to repair it: replacement parts had been hurried in, and production now was higher than ever. In February 1944 the Norwegian Resistance reported that the entire heavy water stock was about to be sent back to Germany.

$$E=mc^2 \cdot E=mc^2 \cdot E=mc^2$$

What to do? It was an excruciating moment, previewing the dilemmas the Allied physicists would face in the decision to use the bomb one year later. Another direct assault wasn't possible, for the Vemork factory was too heavily barricaded. The main train tracks out were heavily guarded as well—there were regular Army troops; SS detachments; auxiliary airfields that would be opened for spotter aircraft.

The sole weak spot for attacking the shipment back to energy was where the train cars with the heavy water from Vemork had to be loaded onto a ferry to cross Lake Tinnsjo, on their way to the Norwegian coast. That was scheduled to take place on February TK, 1944.

If the train was sunk while it was on the ferry, no German divers could bring it up from the lake's depths. But Tinnsjo was also the main crossing to the rest of Norway for the factory workers at Vemork plus their families; it was also a popular tourist crossing. Ordinary families out for the day always were on the ferry.

Whom do you kill for a greater good?

Because of the equation—these powers $E=mc^2$ was offering—the physicists were demanding an awful moral trade-off, greater than should be asked of anyone. Knut Haukelid was one of the Norwegians who had remained behind after the factory raid, living rough on the Hardanger plateau, surviving massive manhunts. By now he was very experienced at the skills needed for sabotage: smuggling himself into a town; working out whom to trust; assembling and testing whatever explosives and timers would be needed. But that wasn't the issue. He had traveled this far, and lived this harshly, to save his countrymen. Now he would be killing them, drowning them in deep cold water.

Norway command to London:

FORMAT AS EXTRACT————REPORTS AS FOLLOWS: . . .
DOUBT IF RESULT OF OPERATION IS WORTH REPRISALS
STOP WE CANNOT DECIDE HOW IMPORTANT THE OPERA-
TIONS ARE STOP PLEASE REPLY THIS EVENING IF POSSIBLE
STOP

London to Norway command:

FORMAT AS EXTRACT————MATTER HAS BEEN CONSID-
ERED STOP IT IS THOUGHT VERY IMPORTANT THAT THE
HEAVY WATER SHALL BE DESTROYED STOP HOPE IT CAN

BE DONE WITHOUT TOO DISASTROUS RESULTS STOP
SEND OUR BEST WISHES FOR SUCCESS IN THE WORK STOP
GREETINGS

The best Haukelid could do was arrange with the Vemork transport engineer that the shipment would only come out on Sunday the 20th, when traffic would be light. (Trade union activity had always been strong in Vemork, and as a result the Resistance had very high membership or at least support in the factory.) Late the Saturday night before, Haukelid arrived with two locals at the berthed ferry. They got on board safely, but when they were hunting for a spot belowdecks to lay their explosives, the night watchman, a young Norwegian, found them. But he knew one of Haukelid's companions, from a local sports club, and quickly nodded his agreement when they gave him their cover story: that Haukelid and the other man, Alf Larsen, had to hide from the Germans, and needed somewhere to store their packages. While the first two men stayed behind talking, Haukelid and Larsen set the charges: right against the front hull, so the explosion would tip the boat forward, lifting the propeller uselessly up in the air, in position to fill with water and sink immediately. It was a half hour before Haukelid was done.

> When I left the watchman, I was not clear
> in my mind as to what I ought to do. ... I
> remembered the fate of the two Norwe-
> gian guards at Vemork, who had been
> sent to a German concentration camp af-
> ter the attack there. I did not want to
> hand over a Norwegian to the Germans.
> But if the watchman disappeared, there
> was danger of the Germans' suspicions
> being aroused next morning.
>
> I contented myself with shaking hands
> with the watchman and thanking him—
> which obviously puzzled him.

Everyone involved was in Haukelid's position. Larsen had been at a dinner party earlier
that evening, where a visiting violinist said that
he'd be taking the boat the next day. Larsen
had tried to say No, he should stay longer in
this beautiful region, the skiing was so excellent. But when the violinist waved that off,
Larsen hadn't been able to insist. A contact at
the factory had told him that his elderly
mother, too, was planning to take the ferry.

The bomb went off at 10:45 A.M.; the boat
was in 1,300 feet of water. The flatcars from
the train broke loose in the sudden tilt, their
doors bursting open. The factory worker's
mother wasn't on board—he'd not let her out
of her house—but the violinist was. There

were fifty-three people on board. Most of the sturdy German guards managed to fight their way off the tumbling ship in time, but many of the women and children were pushed aside. At least twenty-seven of the passengers were caught inside.

A few of the barrels that had been only slightly purified bobbed on the top of the lake, and the passengers who'd managed to get off—including the violinist—grabbed on till a rescue boat came. But the barrels that contained the concentrated heavy water demonstrated, in slow-motion free fall, what they contained. Since the H_2O molecules are composed of a nucleus heavier than ordinary water, the barrels sank as if weighted, swirling around the ferry and its innocent trapped passengers down to the bottom.

$E=mc^2$ · $E=mc^2$ · $E=mc^2$

One year and six months later, in August 1945, 50 pounds of purified Uranium 235, encased within 11,000 pounds of cordite, steel temper, casing, and firing controls, was waiting on a heavy trolley, about to be loaded onto a B-29 on the island of Saipan, six hours' flying time from Japan. Oppenheimer was back in Los Alamos, monitoring this final operation.

If he were a simpler man, he might have been proud. The construction "machine" of researchers and factories and assembly units, which Heisenberg had abortively tried to put together in Germany, had here—on American shores, and under Oppenheimer's direction—finally been achieved. Rivers had been diverted to supply the processing plants and reactors; whole cities had been built to house tens of thousands of workers; a new element had been created through transmutation. It was an immense achievement.

Fermi's first neutron source, the one he'd used in Rome, based on Chadwick's design, could be held on the palm of one hand. The next device Fermi built, scraping together minor funds in New York in 1940, was about the size of a few large filing cabinets. By late 1942, with Oppenheimer overseeing the first substantial U.S. government funding, Fermi had built an enhanced device that filled much of a competition squash court, underneath the stands of the University of Chicago stadium. The final version, constructed two years later, when atomic bomb funding was at full tilt, were the centerpieces of a 300,000-acre site in central Washington, near Hanford. With their supporting structures, they stretched taller than the entire Rome institute where Fermi had begun in 1934. Individuals who were aware of the full history could only stand in front of it in awe.

The plutonium problem had been solved, through mathematicians and explosives experts finding a shape for the ordinary explosives that would smoothly implode the plutonium ball. Regular supplies of the Washington site's output could now the machined for more bombs. The less successful Tennessee factories had also managed to produce a small amount of explosive, and it was Tennessee's total output—almost the complete amount of U235 the United States had—that was being loaded on Saipan.

Heisenberg's work had been blocked. Earlier in 1945, advancing Allied armies in Germany had found entire underground factories, with row upon row of completed jet-powered and even a few rocket-powered aircraft. But the Lake Tinnsjo sinking the previous year had guaranteed that only the barest amount of atomic construction could continue going forward. Even so, Heisenberg had tried to continue. Back in 1942, when funding had looked like it might slow down, he had eagerly explained the possible power of an atomic bomb to a conference of top Nazi administrators, in a quest to get funding back up. Now, even with the war near certain to be lost, he directed that the work be carried on from the small town of Hechingen, where he ended up lodging directly across the street from the home where Einstein's rich uncle had lived— the one who'd supported the family's business efforts, thereby giving Albert the subsidized years to prepare for university entrance.

The equipment lugged from Berlin and Leipzig had been ingeniously installed in a place observation planes wouldn't be able to find. It was put in a cave in an adjacent town, and the cave was in the side of a cliff, and on the top of the cliff was a church—and that was all you would see from the sky. Heisenberg had always been the one for grand gestures. When he'd first conceived of quantum mechanics, one night on a North Sea resort island at age twenty-four, he'd climbed the nearest dune peak and waited there till dawn, copying the Romantic characters from a Caspar David Friedrich painting. Now, in occasional excursions from the cave, he would climb to the highest point in the town, and go into the church, and there in his solitude play Bach with eloquent fury on the organ.

The atomic reactions had gone well beyond the old Leipzig work. By the end, the German researchers had reached about half the rate of nucleus splitting needed for a sustained chain reaction. Heisenberg knew he wouldn't get further. When a U.S. snatch squad did reach him in the Alps, even while Wehrmacht troops were still fighting in adjacent towns, he accepted surrender as if he'd been expecting it.

Heisenberg would be welcomed as a hero in Germany when he was finally released in 1946, while Oppenheimer, even before the war ended, knew his postwar life wouldn't be so simple. He had been a leftist in the late 1930s, and although a Berkeley physics professor might not suffer harm from that, once he was head of Los Alamos the FBI had dug up everything. Then he'd lied about some of the details, in his first interviews with military intelligence. Several important individuals wanted him out, but Groves was protecting him, so in revenge his enemies simply tormented him: all his time as director his phone was tapped, his living quarters bugged, his past friends interrogated, and his trips shadowed. His wife had started drinking, a lot, and although he hadn't yet been attacked, he knew he was open to blackmail: the FBI had followed him on visits to San Francisco, where he'd spent nights with a girlfriend he'd been close to in the past.

More important, he knew of what had happened on Lake Tinnsjo, he knew what was in store in the Pacific. It's common today to believe that the atomic bombing of Japan was justified, on the grounds that the alternative would have been an invasion that had to be much worse. But at the time it was not so clear. The bulk of Japan's army was no threat to American forces: it was sequestered up in China, with American submarines keeping it from crossing to the home islands, and the great weight of Russia's army looming above, able to destroy it once sufficient buildup had occurred. Japan's industry had largely been burned out. Early in 1945, U.S. strategic bombers had been assigned the task of destroying thirty to sixty large and small cities. By August, they had destroyed fifty-eight of them.

Douglas MacArthur, who had run much of the Pacific campaign, didn't expect an invasion would be needed; Admiral Leahy, head of the main Pacific fleet, was adamant that there was no need for an atomic bomb; Curtis LeMay, the head of the strategic bombing force agreed. Even Eisenhower, who'd had no qualms about killing thousands of opponents when it was necessary to safeguard his troops, was strongly hostile to it, as he explained at the time to Henry Stimson, the elderly secretary of war: "I told him I was against it on two counts. First, the Japanese were ready to surrender and it wasn't necessary to hit them with that awful thing. Second, I hated to see our country be the first to use such a weapon. Well . . . the old gentleman got furious. . . ."

The feeling it might not be needed was so strong that there was talk about having demonstrations first, or at least adjusting the phrasing in the surrender demands to make clear that the emperor could remain in place. Oppenheimer had been at many of those meetings: listening intently, arguing—often in a slightly hedged way—for use if it was needed, but supporting the clause about safeguarding the emperor.

These arguments didn't take. Truman's most forceful adviser was Jimmy Byrnes, a man of Lyman S. Briggs's generation, but far less mild in temperament. The ethos Byrnes had been brought up in was that when you fought, you fought with everything you had. He'd been raised in South Carolina in the 1880s, with no father and not a great deal of schooling. Visitors to his state at that time reported their amazement that it was rare on a jury to find twelve men who had all their eyes and ears: South Carolina still had the ethos of a frontier society, and gouging, biting, and knife slashes were the way fights were settled. It was Byrnes who ensured that the clause protecting the emperor—which might mollify Japanese opponents of a settlement—was taken out. There would be no nonsense either about just waiting for the submarine blockade to tighten, or getting advancing Russians to do the dirty work.

Notes from the Presidential "Interim Committee," June 1, 1945: *Mr. Byrnes recommended,* and the Committee *agreed,* that . . . the bomb should be used against Japan as soon as possible; that it be used on a war plant surrounded by workers' homes; and that it be used without prior warning."

Part of Oppenheimer accepted that; part of him—especially when away from Washington—was unsure. But did it matter? He'd helped bring out these powers but now was the least part of it. Oppenheimer's superior, Leslie Groves, was General Groves. Los Alamos was a project of the United States Army. The army built weapons to use them.

The atomic bomb was going to be loaded onto that airplane.

CHAPTER THIRTEEN

8:16 A.M.—Over Japan

Whistling, spinning, the bomb ("an elongated trash can with fins") had taken forty-three seconds to fall from the B-29 that released it. There were small holes around its midpoint where wires had been tugged out of it as it dropped away: that had started the clock switches of its first arming system. More small holes had been drilled farther back on its dark steel casing, in New Mexico, and those took in samples of air as the free fall continued. When it had tumbled to 7,000 feet above the ground, a barometric switch was turned, priming the second arming system.

From the ground the B-29 was just visible as a silvery outline, but the bomb—a bare ten feet long, two and a half feet wide—would have been too small a speck to see. Weak radio signals were being pumped down from the bomb to the Shina Hospital directly below. Some of those radio signals were absorbed in the hospital's walls, but most were bounced back skyward. Sticking out of the bomb's back, near the spinning fins, were a number of whiplike thin radio antennae. Those collected the returning radio signals, and used the time lag each took to return as a way of measuring the height remaining to the ground. At 1,900 feet the last rebounded radio signal arrived. John von Neumann and others had calculated that a bomb exploding much higher would dissipate much of its heat in the open air; exploding much lower, it would dig a huge crater in the ground. At just under 2,000 feet the height would be ideal. An electric impulse lit cordite sacs, producing a conventional artillery blast. A small part of the total purified uranium was now pushed forward down a gun barrel that was actually inside the bomb. In the early planning this gun had been a very heavy device, being simply a copy of large U.S. Navy weapons. Only after several months had one of Oppenheimer's men realized that navy guns were so heavy because they had to survive the recoil of shot after shot. Here, of course, it wouldn't matter: this gun was only going to be fired once. Instead of weighing 5,000 pounds it was machined to weigh barely a fifth of that. The uranium segment traveled about four feet within the thinned gun barrel, and then it impacted the remaining bulk of the uranium. Nowhere on Earth had a ball of several dozen pounds of such purified uranium ever been accumulated. There were a number of stray neutrons loose inside it, and although the uranium atoms were densely protected by their outer flurries of electrons, the escaped neutrons, having no electrical charge, weren't affected by the electrons. They flew through the outer electron barrier—as we saw, like a probe skimming past the planets down toward our sun—and while many of them flew straight on out the other side, a few were on a collision course for the speck of a nucleus far down at the center. That nucleus normally blocked outside particles from entering, for it was seething with positively charged protons. But neutrons have no electric charge. To the protons, at least from a distance, they are "invisible." The arriving neutrons pushed in to the nucleus, overbalancing it; making it jostle and wobble. The uranium atoms mined on Earth were each over 4.5 billion years old. Only a very powerful force, before the Earth was formed, had been able to squeeze their electrically crackling protons together. Once that uranium had been formed, the strong nuclear force had acted, gluelike, to hold the protons in place over all that long span: while the Earth cooled, and continents formed; as America separated from Europe, and the North Atlantic Ocean slowly filled; as volcanic bursts widened on the other side of the globe, forming what would become Japan. A single extra neutron unbalanced that stability now. Once the wobbling in the nucleus was enough to break the strong force glue, then the ordinary electricity of the protons was available to force them apart. A single nucleus doesn't weigh much, and the fragmentary section of one weighs even less. Its speeding impact into the other parts of the uranium didn't heat it up much. But the density of uranium was enough that a chain reaction started, and soon there weren't just two speeding fragments of uranium nuclei, there were four, then eight, then sixteen, and so on. Mass was "disappearing" within the atoms, and coming out as the energy of speeding nuclei fragments. $E=mc^2$ was now under way. The entire sequence of multiplying releases was finished in just under one-millionth of a second. The bomb was still suspended in the humid morning air with a faint layering of condensation on its outer surface, for it had been up in the cold air of 31,000 feet just forty-three seconds earlier, and now, 1,900 feet over the hospital, it was a balmy 80°F. the bomb fell downward just an additional fraction of an inch in the time of most of the reaction; from the outside there would only be the first odd bucklings of its steel surface to suggest what was going on inside. The chain reaction went through eighty "generations" of doubling before it ended. By the last few of those, the segments of broken uranium nuclei were so abundant, and moving so fast, that they started heating up the metal around them. The last few doublings were the crucial ones. Imagine you have a pond in your garden, with a lily plant floating on it that doubles in size every day. In eighty days the lily entirely covers the pond. On which day is half the pond still uncovered, open to the sun and outside air? It's the seventy-ninth day. From this point on, all the action of the $E=mc^2$ reaction was over. No more mass was "disappearing"; no more fresh energy appeared. The energy in the movement of those nuclei was simply being transformed to heat energy—just as rubbing your hands together will make your palms warm up. But the uranium fragments were rubbing against resting metal at immense speed, due to the multiplication by c^2. They soon were traveling at a substantial fraction of the speed of light. The rubbing and battering made the metals inside the bomb begin to warm. They had started at near body temperature—98.6°F or 37°C—and then they reached water's boiling temperature—

212°F or 100°C—and then that of lead—560°C. But the generations of chain reaction doubling had gone on, as yet more uranium atoms had been splitting, so it reached 5,000°C (the surface of the sun) and then several million degrees (the temperature of the center of the sun) and then it kept on rising. For a brief period, in the center of the suspended bomb, conditions similar to those in the early moments of the creation of the universe were produced. The heat moves out. It goes through the steel tamping around the uranium, and just as easily through what had been the several-thousand-pound massive casing of the bomb, but then it pauses. Entities as hot as that explosion have energy that must be released. It starts pushing X rays out of itself, a very large number of them, angling some of them up, and some to the side, and the rest in a wide stretching arc toward the ground. The explosion is hovering; the fragments are trying to cool themselves off. It remains that way, pouring out a large part of its energy. Then, after 1/10,000th of a second, when the X ray spraying is over, the heat ball resumes its outward spread. Only now does the central eruption become visible. Ordinary light photons could not push through the X ray sprays; only the glows on the outside of the sprays would have been seen. When the full flash appears, it's as if a rip in the sky has opened. An object resembling one of the giant suns from a distant part of our galaxy now appears. It fills several hundred times more of the sky than Earth's ordinary sun. The unearthly object burns at full power for about one-half of a second, then begins to fade away, taking two or three seconds to empty itself out. This "emptying" is accomplished, in large part, by spraying heat energy outward. Fires begin, seemingly instantaneously; skin explodes off, hanging in great sheets from the bodies of everyone below. The first of the tens of thousands of deaths in Hiroshima begin. At least a third of the energy from the chain reactions comes out in this flash. The rest now follows soon behind. The strange object's heat pushes on ordinary air, accelerating it to speeds that have never occurred here before, unless at some time in the distant past a large meteor or comet arrived. It travels several times faster than any hurricane could achieve—so fast, in fact, that it's silent, for it outruns any sound its immense force might make. After it there's a second air pulse, a little slower; after that the atmosphere sloshes backward, to fill up the gap pushed out. This briefly lowers the air density to virtually zero. Far enough from the blast, life-forms that have survived will now begin to explode outward, having been exposed—briefly—to the vacuum of outer space. A small amount of the heat that was produced can't move forward at all. It remains behind, hovering quite close to where the fuses and antennae and cordite had been. In a few seconds it begins to rise. It swells as it goes, and at sufficient height it spreads out. And when that great mushroom cloud appeared, $E=mc^2$'s first work on planet Earth was done.

Part Five . . .

. . . Till the End of Time

CHAPTER FOURTEEN

The Fires of the Sun

The flash of light from the explosion over Hiroshima in 1945 reached the orbit of the moon. Some bounced back to Earth; much of the rest continued onward, traveling all the way to the sun, and then indefinitely beyond. The glare would have been viewable from Jupiter.

In the perspective of the galaxy, it was the most insignificant flicker.

Our sun, alone, explodes the equivalent of 16,000,000,000 such bombs every second. For $E=mc^2$ does not apply just on Earth. All the scrambling commandos and anxious scientists and cold-eyed bureaucrats: all that is but a drop, the slightest added whisper, in the enormous powerful onrushing of the equation.

Einstein and other physicists had long recognized this: it was just a quirk that the accelerated technology and pressures of wartime had led to the equation's first applications being focused on weaponry. In this section of the book we switch to those wider views; lifting away from earthly technology, and showing how the equation's sway extends throughout the universe: controlling everything from how the first stars ignited, to how life will end.

$E=mc^2$ · $E=mc^2$ · $E=mc^2$

Ever since the discovery of radioactivity in the 1890s, researchers had suspected that uranium or a similar fuel might be operating in the wider universe, and in particular, in our sun to keep it burning. Something that powerful was needed, because Darwin's insights as well as findings in geology had shown that Earth must have been in existence—and warmed by the sun—for billions of years. Coal or other conventional fuels would not be strong enough to do that.

Unfortunately, though, astronomers couldn't find any signs of uranium in the sun. Every element gives off a distinctive visual signal, and the optical device called the spectroscope (it breaks apart the "spectrum") allows them to be identified. But point a spectroscope at the sun, and the signals are clear: there is no uranium or thorium or other known radioactivity glowing element up there.

What did seem to leap out, in readings from distant stars as well as our sun, was that there was always iron inside them: lots and lots of metallic bulky iron. By the time Einstein was finally able to leave the patent office, in 1909, the best evidence was that the sun was about 66 percent pure iron.

This was a disheartening result. Uranium could pour out energy in accord with $E=mc^2$, because the uranium nucleus is so large and overstuffed that it barely holds together. Iron is different. Its nucleus is one of the most perfect and most stable imaginable. A sphere made of iron—even if it was molten or gaseous or ionized iron—could not pour out heat for thousands of millions of years.

Suddenly the vision of using $E=mc^2$ and related equations to explain the whole universe was blocked. Astronomers could just look to the top of the atmosphere, to the great spaces and waiting suns beyond us, and wonder.

$E=mc^2$ · $E=mc^2$ · $E=mc^2$

The individual who broke that barrier— letting $E=mc^2$ slip the surly bonds of Earth— was a young Englishwoman named Cecilia

Payne, who loved seeing how far her mind could take her. Unfortunately, the first teachers she found at Cambridge when she entered in 1919 had no interest in such explorations. She switched majors, and then switched again, which led to her reading up on astronomy, and when Payne started *anything*, the effects were impressive. She terrified the night assistant at the university's telescope her first night there, after she'd been reading for only a few days. He "fled down the stairs," she recalled, "gasping: . . . 'There's a woman out there asking questions.'" But she wasn't put off, and a few weeks later she described another such incident: "I bicycled up to the Solar Physics Observatory with a question in my mind. I found a young man, his fair hair tumbling over his eyes, sitting astride the roof of one of the buildings, repairing it. 'I have come to ask,' I shouted up at him, 'why the Stark effect is not observed in stellar spectra.'"

This time her subject did not flee. He was an astronomer himself, Edward Milne, and they became friends. Payne tried to pull her arts student friends into her astronomical excitements, and even though they might not have understood much of what she was saying, she was the sort of person others like being around. Her rooms at Newham College were almost always crowded. A friend wrote:

". . . when safely lying on her back on the floor (she despises armchairs), she will talk of all things under the sun, from ethics to a new theory of making cocoa."

Rutherford was teaching at Cambridge by then, but didn't know what to do with Payne. With men he was bluff and friendly, but with women he was bluff and pretty much a thug. He was cruel to her at lectures, trying to get all the male students to laugh at this one female in their midst. It didn't stop her from going— she could hold her own with his best students in tutorials—but even forty years later, retired from her professorship at Harvard, she remembered the rows of braying young men, nervously trying to do what their teacher expected of them.

But also at the university was Arthur Eddington, a quiet Quaker who was happy to take her on as a tutorial student. Although his reserve never lifted—tea with students was always in the presence of his elderly unmarried sister—the twenty-year-old Payne picked up Eddington's barely stated awe at the potential power of pure thought.

He liked to show how creatures who lived on a planet entirely shrouded in cloud would be able to deduce the main features of the unseen universe above them. There would have to be glowing spheres up in space, he imagined them reasoning, for a ball of vaporized elements sufficiently large, and sufficiently dense would compress the elements inside it to start a nuclear reaction that would make it light up—it would be a sun. These glowing spheres would be dense enough to pull planets swinging around them. If the beings on the mythical planet ever did find that a sudden wind had blown an opening in their clouds, when they looked up they'd see a universe of glowing stars, with circling planets, just as they'd expected.

It was exhilarating to think that someone on Earth might solve the problem of how to deal with all the iron in the sun, and so be able actually to work out Eddington's vision. When Eddington first assigned Payne a problem on stellar interiors, which might at least start to achieve this, "the problem haunted me day and night. I recall a vivid dream that I was at the center of the giant star Betelgeuse, and that, as seen from there, the solution was perfectly plain; but it did not seem so in the light of day."

But even with this kind man's backing, a woman couldn't do graduate work in this field in England, so she went to Harvard, and there blossomed even more. She switched from her heavy woolen clothing to the lighter fashions of 1920s America; she found a thesis adviser, Harlow Shapley, an up-and-coming astrophysicist; she loved the liberty she found in the student dorms, and the fresh topics in the university seminars. She was bursting with enthusiasm.

And that could have blocked everything. Raw enthusiasm is dangerous for young researchers. If you're excited by a new field—keen to join in with what your professors and fellow students are doing—that usually means you'll be trying to fit in with their approaches. But students whose work stands out usually have had some reason to avoid this, and keep a critical distance. Einstein didn't especially respect his Zurich professors: most, he thought, were drudges, who never questioned the foundations of their teaching. Faraday couldn't be content with explanations that left out the inner feelings of his religion; Lavoisier was offended at the vague, inexact chemistry handed down by his predecessors. For Payne, some of her needed distance came from getting to know her fun fellow Ivy League students a little better.

Shortly after arriving: "I expressed to a friend that I liked one of the other girls in the House where I lived at Radcliffe College. She was shocked: 'But she's a Jew!' was her comment. This frankly puzzled me. . . . I found the same attitude towards those of African descent."

She also got a glimpse of what was going on in the back rooms at the Observatory. In 1923, the word *computer* did not mean an electrical machine. It meant people whose sole job was to compute. At Harvard, it was applied to ranks of slump-shouldered spinsters in those back rooms. A few of them had once had first-rate scientific talent ("I always wanted to learn the calculus," one said, "but the director did not wish it"), but that was usually long since crushed out of them, as they were kept busy measuring star locations, or cataloging volumes of previous results. If they got married they could be fired; if they complained of their low salaries, they would be fired as well.

Lise Meitner had had her problems in getting started in research in Berlin, but there was nothing like this desolate, life-crushing sexism. A few of the Harvard "computers," in several decades of bent-back work, succeeded in measuring over 300,000 spectral lines. But what it meant, or how it fit in with the latest developments in physics, was not for them to understand.

Payne was not going to be pushed into their ranks. Spectroscope readings can be ambiguous where they overlap. Payne began to wonder how much the way her professors broke them apart depended on what they already had in mind. For example, let the reader note the following letters very well, and then try to read them:

```
n   o   t   e
v   e   r   y
o   n   e   w
i   l   l   g
e   t   i   t
```

It's not easy. But if you start reading it instead as "Not everyone . . ." then it leaps out. What Cecilia Payne decided on, there in 1920s Boston, was a Ph.D. project that would let her confirm and further develop a new theory about how to build up spectroscope interpretations. Her work was more complicated than our example above, for spectroscope lines from the sun will always include fragments of several elements; there are distortions from the great temperature as well.

An analogy can show what Payne did. For astronomers convinced there's going to be lots of iron in the sun (which seemed fair for there was so much iron on Earth and in asteroids), there'd be only one way to read an ambiguous string of lines from a spectroscope. If they came out, for example, as:

t h e y s a i d i r o n a g a i e n

You'd parse it to read:

t h e y s a i d **I r o n** a g a i e n

and there'd be no need to worry too much about the odd spelling of *agaien*. The extra *e* could be a fault in the spectroscope, or some odd reaction on the sun, or just a fragment that was slipped in from some other element. There's always something that doesn't fit. But Payne kept an open mind. What if it was really trying to communicate:

t **H** e **y** s a i **d i r o** n a **g** a i **e n**

She went through the spectroscope lines over and over again, checking for these ambiguities. Everyone had boosted the lines one way, to make it read as if they were for iron. But it wasn't too much of a stretch to boost them differently,

So they read iron, not hydrogen.

Even before Payne finished her doctorate, her results bean to spread in gossip among astrophysicists. While the old explanation of the spectroscope data had been that the sun was two-thirds iron or more, this young woman's interpretation was that it was over 90 percent *hydrogen,* with most of the rest being the nearly as lightweight helium. If she was right, it would change what was understood about how stars burn. Iron is so stable that no one could imagine it transformed through $E=mc^2$ to generate heat in our Sun. but who knew what hydrogen might do?

The old guard knew. Hydrogen would do nothing. It wasn't there, it couldn't be there; their careers—all their detailed calculations, and the power and patronage that stemmed from it—depended on iron being what was in the sun. After all, hadn't this female only picked up the spectroscopic lines from the sun's outer atmosphere, rather than its deep interior? Maybe her readings were simply confused by the temperature shifts. Her thesis adviser declared her wrong, and then *his* old thesis adviser, the imperious Henry Norris Russell, declared her wrong, and against him there was very little recourse. Russell was an exceptionally pompous man, who would never accept he could be wrong—and he also controlled most grants and job appointments in astronomy on the East Coast.

For a while Payne tried to fight it anyway: repeating her evidence; showing the way her hydrogen interpretation was just as plausible in the spectroscope lines as the iron ones; even more, the way new insights—the latest in European theoretical physics—were suggesting a way hydrogen really could power the sun. It didn't matter. She even tried reaching out to Eddington, but he withdrew, possibly out of conviction; possibly out of caution before Russell; or possibly just from a middle-aged bachelor's fear of a young woman turning to him with emotion. Her friend from her student days at the Cambridge Solar Physics Observatory, the young fair-haired Edward Milne, was by now an established astronomer, and did try to help, but he didn't have enough power. Letters were exchanged between Payne and Russell, but if she wanted to get her research accepted she'd have to recant. In her own published thesis she had to insert the humiliating line: "The enormous abundance of hydrogen . . . is almost certainly not real."

A few years later, though, and the full power of Payne's work became clear, for independent research by other teams backed her spectroscope reinterpretations. She was vindicated, and her professors were shown to have been wrong.

$E=mc^2$ · $E=mc^2$ · $E=mc^2$

Although Payne's teachers never really apologized, and tried to hold down her career as long as they could, the way was now open to applying $E=mc^2$ to explain the fires of the sun. She had shown that the right fuel was floating up in space; that the sun and all the stars we see actually are great $E=mc^2$ pumping stations. They seem to squeeze hydrogen mass entirely out of existence. But in fact they're simply squeezing it along the equals sign of the equation, so that what had appeared as mass, now bursts into the form of billowing, explosive energy. Several researchers made starts on the details, but the main work was done by Hans Bethe, the same man who later co-wrote that 1943 memo to Oppenheimer about the on-going German threat.

Down on Earth, the few hydrogen atoms in our atmosphere just fly past each other. Even if crushed under a mountain of rock, they won't really stick to each other. But trapped near the center of the sun, under thousands of miles of weighty substance overhead, hydrogen nuclei can be squeezed close enough together that they will, in time, join together, to become the element helium.

If this were all that happened, it wouldn't be very important. But each time four nuclei of hydrogen get squeezed together, Bethe and the others now showed that they follow the potent, subatomic arithmetic of the sort Meitner and her nephew Frisch had worked on that afternoon in the Swedish snow. The mass of the four hydrogen nuclei can be written as 1+1+1+1. But when they join together as helium, their sum is not equal to 4! Measure a helium nucleus very carefully, and it's about 0.7 percent less, or just 3.993. That missing 0.7 percent comes out as roaring energy.

It seems like an insignificant fraction, but the sun is many thousands times the size of Earth, and the hydrogen in this tremendous volume is available as fuel. The bomb over Japan had destroyed an entire city, simply from sucking several ounces of uranium out of existence, and transforming it into glowing energy. The reason the sun is so much more powerful is that it pumps 4 million *tons* of hydrogen into pure energy each second. One could see our sun's explosions clearly from the star Alpha Centauri, separated from us by 24 trillion miles of space; and from unimagined planets around stars far along the spiral arm of our galaxy as well.

The sun did that much pumping yesterday when you woke up—4 million tons of hydrogen "squeezed" along Einstein's 1905 equation from the mass side to the energy side, getting multiplied by the huge figure of c^2— and it was pouring out that much energy at dawn over Paris five centuries ago, and when Mohammed first took refuge in Medina, and when the First Dynasty was established in China. Energy from millions of disappearing tons was roaring overhead each second when the dinosaurs lived: Earth has been nurtured, and warmed, and protected, by this same raging fire as long as it has been in orbit.

CHAPTER FIFTEEN

Creating the Earth

Cecilia Payne's work had helped show that our sun and all the other stars in the heavens are great $E=mc^2$ pumping stations. But on its own, hydrogen-burning could easily have led to a sterile, dead universe. Early in the universe's history, there would have been great blazings as the hydrogen stars created their helium. But the original hydrogen fuel would eventually have used itself up, and the fires explained by $E=mc^2$ would have gradually died down, leaving only giant floating ash heaps of used helium. Nothing else would have ever been created.

To create the universe we do know, there had to be some device for building the carbon and oxygen and silicon and all the other elements that planets and life depend on. These elements are larger and more complex than what a simple hydrogen-to-helium combustion machine could ever produce.

Payne had been independent enough to challenge the consensus that stars were made of iron, and this had allowed the first stage of insight: showing that there actually was enough hydrogen in the stars above our atmosphere to allow the energy-spraying sequence of 1+1+1+1 not quite 4.00 to occur, thereby sustaining their fires. But with the production of helium, there it stopped. Who would be cockily independent enough to take it further, and show how $E=mc^2$ could operate to create the elements of our planet and daily life as well?

$E=mc^2$ · $E=mc^2$ · $E=mc^2$

In 1923, when Payne arrived at Harvard, a seven-year-old Yorkshire lad was found by his local truant officer to have been spending most of the past year at the local cinema. Even though young Fred Hoyle explained most forcefully that it had been good for his education—he'd taught himself to read by following the subtitles—he was forced, against his will, back to school. It would be this young boy's work that ultimately solved the next major step in how the sun burns.

One year after Hoyle returned to school, his class was assigned to collect wildflowers. Back in the classroom the teacher read out the list of flowers, describing one as having five petals. Hoyle examined the sample he'd collected, now in his hand. It had six petals. This was curious. If it had been a petal less than described, that would have been understandable, he might have torn one off in carrying it. But how could there be more? He was puzzling over it, and vaguely heard a strident voice, and then: "The blow was delivered flat-handed across the ear," he wrote, ". . . the one in which I was to become deaf in later life. Since, moreover, I wasn't expecting it at all, I had no opportunity to flinch by the half inch or so that would have reduced the impulsive pressure on my drum and middle ear."

It took a few minutes for Hoyle to recover, but then he left the school, and back at home explained to his mother what had happened: "I pointed out, I'd given the school system a tryout over three years, and, if you didn't know something was no good after three years, what did you know?"

His mother pretty much agreed, and so did his father, who had survived two years as a machine gunner on the Western Front by disobeying the less than brilliant orders from his upper-class officers to test-fire his guns at ten-minute intervals (which would have given his squad's exact location to German assault teams). Fred Hoyle got yet another year off. "Each morning, I ate breakfast and started off from home, just as if I were going to school. But it was to the factories and workshops of Bingley that I went. There were mills with clacking and thundering looms. There were blacksmiths and carpenters. . . . Everybody seemed amused to answer my questions."

In time he was railroaded back to school, where a few kind teachers saw his talent, and helped with scholarships. He ended up studying mathematics and then astrophysics at the University of Cambridge, and he did so well that the intensely private Paul Dirac took him on as a student, which was unheard of, and he even had tea with Payne's old supervisor Eddington—though as there were rumors of some sort of intellectual "disgrace" she had run into at Harvard, Payne's name was now barely mentioned. (History had been rewritten: Henry Norris Russell and the others now implied they'd "always" known that plenty of hydrogen was available in the sun.)

The problem of how stars manage to use helium as a further fuel in the giant $E=mc^2$ pumps, however, hadn't gone much further than where Payne's work and the direct follow-ups had left it in the 1920s. the 20-million-degree heat at the center of our sun was able, barely, to squeeze the positive charges of four hydrogen nuclei together to make helium. To squeeze together those helium nuclei in a burning process to get larger elements, you'd need to get higher temperatures. But the universe was well surveyed.

Where could you find something hotter than the center of a star?

Hoyle's habit of putting things together in his own way now came to the fore. At the start of World War II he joined a radar research group, and in December 1944, after an information-sharing mission to the United States, he ended up waiting in Montreal for a rare flight back across the Atlantic.

He wandered around the city and beyond, visiting the British research group at Chalk River (about 100 miles from Ottawa). Although nobody told him anything official about the Manhattan Project, from the faces he saw there—including several whose work he'd known at Cambridge before the war—he neatly deduced the basic stages of the top-secret project still going on at Los Alamos.

The easiest way to accumulate the raw material for a bomb, he already knew from reading accounts published before the war, was by cooking up plutonium in a reactor. He also

knew that Britain had not tried building reactors. That meant, he concluded, that the specialists must have found some unsuspected
problem with the plutonium route; probably
with getting the ignition to operate fast
enough. Now, though, seeing the specialists in
Canada, including experts in the mathematics
of explosions, he realized it must have been
overcome.

Oppenheimer and Groves had barbed wire
and armed guards and layers of security officers around the plutonium detonation group at
Los Alamos. But that was no protection
against a man who'd managed to outwit the
stern educational establishment of village
Yorkshire. By the time he was finally assigned a
seat on a flight back, Hoyle had deduced what
Oppenheimer's hundreds of specialists had
proven. A substance such as plutonium that
won't full explode on its own will certainly
tear apart its own atoms if it's squeezed inward
abruptly enough. Implosion raises the pressure and temperature enough to do that.

Everyone had thought of implosion as intensely localized; suitable only for plutonium
spheres a few inches across. But why did it
have to stay so small? Implosion was a powerful technique on Earth. Hoyle was used to following his thoughts anywhere. Why couldn't
it apply in the stars?

If a star ever imploded, it too would get hotter. Instead of being 20 million degrees, its center could reach—as Hoyle quickly computed—100 million degrees. That would be enough to squeeze even the larger nuclei of more massive elements together. Helium could be squeezed to create carbon. If the implosion went further, the star would get even hotter, and yet heavier nuclei would be created: oxygen, silicon, and sulfur, and the rest.

It all depended on a star's actually undergoing this inner collapse, but Hoyle realized there was a plausible reason this should happen. When a star was still at the relatively cool 20 million degrees, and capable only of burning hydrogen, the helium that was produced would build up like ash in a fireplace. When all the hydrogens was used up, that ash wouldn't be able to burn. The upper reaches of the star would no longer be pushed outward by the fires within. They would come crashing inward—just as in the Los Alamos bomb.

When a star implodes inward, that would raise its temperature to the 100 million degrees that is enough to ignite the helium ash. When *that* helium is used up, "useless" carbon ash accumulates and the next stage occurs. The carbon can't burn at 100 million degrees, so yet a further level of the star crashes down. The temperature gets higher, and the cycles repeat. It's as if a multifloor building were slowly collapsing, as the struts holding up one floor after another suddenly buckle and break. $E=mc^2$ is central, for each level of burning— first the hydrogen, then the helium, then the carbon—gets its power from the conversion of mass into energy.

There were more details to come next, many contributed by Hoyle himself, but the idea taken from the atomic bomb had been central in solving the problem. Hoyle had simply switched the implosion process from the few pounds of plutonium laboriously collected on Earth, to a sphere of ultraboiling gas—a star—thousands of miles wide, at immense distances away in space. He'd seen how stars can cook up the elements of life. In time it became clear that when the larger of these stars used up their last possible fuel, they'd have to break apart. Everything they had made would then pour out.

$$E=mc^2 \cdot E=mc^2 \cdot E=mc^2$$

We tend to think of our planet as old, but when it was newly formed the heavens were already ancient; full of millions of these exploded giants. Their eruptions flung out silicon, and iron and even oxygen, to make the substance of Earth.

Heavier, unstable elements such as uranium and thorium were created in the ancient stars' explosions as well, and when these elements floated over, becoming incorporated in the deep body of Earth, their continued explosions shot fragments of protons and other atomic parts at high velocity into the surrounding rock.

Along with the initial heat left over from the impacts of Earth's creation, the radioactive blastings from the uranium and similar heavy elements have kept our planet's depths from cooling. The repeated multitudes of $E=mc^2$ bursts from those atoms helped produce a churning heat underneath the surface, enough to make the thin continents on top roll forward, so shaping the surface of Earth.

In some places, sections of the thin crust were pushed crumpling into each other, producing the lifted ripples we call the Alps, Himalayas, or Andes. In other places, the churning heat pulled open gaps that we know by such names as San Francisco Bay, the Red Sea, and the Atlantic. These made excellent collection basins for the hydrogen that had also landed, and when that combined with oxygen, the result was oceans of sloshing water. Iron deep inside the planet sloshed in its own more stately fashion, driven by the twenty-four-hour spinning of the whole globe around its axis. That sent up invisibly streaking magnetic lines, of exactly the sort Michael Faraday described, and reproduced, in the basement of London's Royal Institution 4 billion years later. The result was an invisible network of magnetic force lines, far overhead, helping shield the self-assembling carbon molecules on the surface from some of the worst of the spraying radiation from outer space.

Volcanoes exploded upward—powered by the constant $E=mc^2$ derived heat beneath—and that led to something of a continuous conveyer belt from deep underground. Key trace elements were pushed up into the air, helping produce our fertile soil; great clouds of carbon dioxide were carried upward as well, creating a greenhouse effect in the young planet's atmosphere, and further ensuring the surface warmth needed for life. Where the frictional heat generated by the atoms blasting apart in accord with $E=mc^2$ was especially concentrated, deep-sea volcanoes could push up even through thousands of feet of cold ocean water—which is how the Hawaiian islands lifted above the Pacific waves.

Fast-forward several billion years, and mobile chunks of carbon atoms emerged (in other words, us!) to wade through low-flying clouds of star-created oxygen, stir caffeine-dense liquids of Big Bang hydrogen atoms, and read about how they came to exist. For we live on a planet where $E=mc^2$ is constantly at work in the technology around us.

Atomic bombs were the first direct application. At the start there were just a handful, laboriously created in the labs of the Manhattan Project, but soon there were many more, as a great infrastructure of factories and scholarship and research institutes became established after Hiroshima. Several hundred atomic or hydrogen bombs were built and ready by the end of the 1950s; today, even well after the Cold War, there are more than 12,000. To create them there were hundreds of open-air tests over the years, spraying immense gushes of radioactive particles into the stratosphere, there to float to every location on the planet; becoming a part of the bodies of every person alive.

Nuclear submarines were created, with radioactively exploding uranium sequestered inside; pouring out heat that spun the turbines. They were fearsome weapons, yet thereby allowed a curious stability in the most dangerous phases of the Cold War. The previous generations of submarines, from World War II, had been unable to spend much time at battle stations. Cruising on the surface, World War II submarines might just manage to travel at the 12 mph of a person on a bicycle; taking the safer route, underwater, they moved at the 4 mph of a person walking. Once they'd crossed half the North Atlantic or the Pacific, they'd used up so much fuel that they quite soon had to engage in difficult wartime refueling, or turn around and trundle back. With nuclear-powered engines, it was different. Russian and American submarines could get into firing range, and then stay there for weeks or months on end—a dangerous game, but one at least making the other side very cautious about any moves that might provoke these hidden vessels to launch their missiles.

On land, huge electricity-generating stations were built, using the high-speed frictional heat of $E=mc^2$ to power up generating turbines. It's not the most sensible of energy choices, for even nonnuclear explosions at the generating stations can be pretty terrifying—and nothing deters corporate financial officers as much as the phrase *unlimited liability*; the radioactive walls and radioactive cement base and radioactive residual fuel from every such generator are a lot of liability to be disposed. In France, however, the government assumes those charges, and doesn't allow court cases against the industry: about 80 percent of the country's electricity is nuclear. When the Eiffel Tower is lit at night, the electricity comes from a slower reenactment of the exploding ancient uranium atoms that took place over Hiroshima.

$E=mc^2$ continues at work in ordinary houses. In the smoke detectors screwed tight to the kitchen ceiling, there's usually a sample of radioactive americium inside. The detector gets enough power by sucking mass out of that americium and using it as energy—in exact accord with the equation—that it can generate a smoke-sensitive charged beam, and keep on doing it for months or years on end.

The red-glowing exit signs in shopping malls and movie theaters directly depend on $E=mc^2$ as well. These signs can't rely on ordinary light sources, because they'd fail if the electricity went out in a fire. Instead, radioactive tritium is sealed inside. The signs contain enough fragile tritium nuclei that mass is constantly lost, and usefully glowing energy comes out instead.

In hospitals, medical diagnostics constantly harness the equation. In the powerful imaging devices known as PET scans (Positron Emission Tomography), patients breathe radioactive oxygen isotopes. The nuclei of those atoms shatter apart, and streaks of energy coming from the destroyed mass are recorded as they emerge at extremely high speed from the body. The result is pinpointed readouts on tumors, blood flow, or drug take-ups inside the body—the workings of Prozac in the brain, for example, were detailed this way. In radiation treatment for cancer therapy, milligram quantities of substances such as radioactive cobalt are aimed at tumors. As the cobalt nuclei break apart, mass ounce again is seemingly torn out of the existence, and the resultant energy is aimed with enough power to destroy cancerous DNA.

Outside the window of passenger jets, unstable varieties of carbon are being formed, created by incoming cosmic rays, some of which come from distant reaches of the galaxy. We've been breathing in the stuff all our life. Hold a sufficiently sensitive Geiger counter over your hand, and it registers the telltale clicks. (What it's actually doing is "listening" to tiny miniatures of Einstein's 1905 equation. Every click of the Geiger counter is a mark that one or more operations of $E=mc^2$ has taken place, as the unstable nucleus of that new carbon atom plops out the extra neutron it gained high in the air.) But when we stop breathing—or when a tree dies, or a plant stops growing—no more fresh carbon is coming in. The clicks slowly die away.

This unstable carbon is the famous C-14. It's a clock, and its use has revolutionized archeology. Using carbon dating, labs could prove that the Turin Shroud was a medieval forgery, as some of the carbon in its flax had been running down since the fourteenth century, but not earlier. Carbon fragments could be collected from the Lascaux caverns, and Indian burial mounds, and Mayan pyramids, and early Cro-Magnon sites, and for the first time be used to date them accurately as well.

Soaring even higher, the satellites of the U.S. Defense Department's GPS navigation system create a constantly swirling tessellation beyond the atmosphere. The signals they beam down are constantly shifted out of sync by the time-distorting effects of relativity, as we saw in Chapter 7, and just as steadily have to be fixed, by programmers who adapted Einstein's insights to correct for the drift that would otherwise create. And finally, perched most distant of all, is the exploding sphere of our sun, using the boomingly magnifying power of c^2 to warm our planet, as it has done for all the billions of years needed for this life-dense vista to evolve.

CHAPTER SIXTEEN

A Brahmin Lifts His Eyes
Unto the Sky

Even though the sun is vast, it can't keep on burning forever. Heating the entire solar system takes immense amounts of fuel, even for a furnace that pumps material directly along the equals sign of $E=mc^2$. The sun's mass is now 2,000,000,000,000,000,000,000,000,000 tons, but it consumes 348,000,000,000,000 tons of its own mass as hydrogen fuel to keep the multimegaton blasts going each day. In a further 5 billion years, the most easily available portions of that fuel will be gone.

When that happens, and all that remains at the center is helium "ash," the reactions in our sun will start shifting upward a little bit, as fuel closer to the surface starts being pumped through $E=mc^2$. The outer layers of the sun will expand, and cool down just slightly enough to glow red. The sun will keep on expanding, and keep on glowing, until it reaches Mercury's orbit. That planet's rock surface will have already melted; fragments that are left will now be absorbed in the flames. Then, a few tens of millions of years later, our red-giant sun will reach the orbit of the planet Venus, and absorb it as well. But what will happen next?

Some say the world will end in fire,
Some say in ice.

Robert Frost published that in 1923, when he was pretending to be an apple farmer in Vermont. But he'd written the first draft when he'd been on the faculty at Amherst, and so had a good deal of time to read. Most science writers of the time had settled on the image, popular from Buffon's time through the late Victorian period, of a great cooling down of the universe. But others contrasted that with earlier apocalyptic images from Revelations, where fire and outpourings took over at the end.

What will happen to Earth is actually both. Any beings left alive on the surface of the Earth in the year A.D. 5 billion will see the sun get larger and larger until it fills about half the daytime sky. The oceans will boil away, and surface rocks will melt. Possibly life could migrate to other planets, or survive in deep tunnels, using technology unimagined now; maybe our planet will have long been barren when the emptying sun fills up the sky.

The sun will hold at that great size for about another billion years, as the helium ash left inside takes over the main burning: still seeming to pump mass out of existence; still replacing it with fiery glowing energy. Then it'll shrink, as the supporting struts of that glowing energy become too weak. In time so much fuel will have been emptied out of the sun that the burning is no longer steady.

This is what will bring in the ice. As fuel pockets inside run low, the sun's surface will sink inward; shortly after, as other dispersed fuel sources get tapped, the energy output will roar higher again, and the surface of the sun will whip upward. Sonic booms are produced each time, but these are nothing like the brief crack of a single plane passing the sound barrier. At this stage, six billion years into our future, it's the final boom of the Titans.

Enough mass is blown away at each bounce upward, that within just a few hundred thousand years, there will be much less of our sun than before. What's left will be too weak to possess the same gravitational field it had before. If the Earth hasn't already been absorbed by the expanding sun, then—after 11 billion years of steady orbit—the sun's grip will let the planets go. The solar system breaks up, and Earth flies away.

$$E=mc^2 \cdot E=mc^2 \cdot E=mc^2$$

One of the key insights into what happens next—and within which $E=mc^2$ is once again crucial—was first made by Subrahmanyan Chandrasekhar, a leader in twentieth-century astrophysics, whose career spanned almost sixty years. The discovery came when he was just nineteen, in the hot summer of 1930. The British Empire was in its dying days, but Chandra (the name he usually went by) was still within its dominion, and en route from Bombay to England, where he was taking up graduate studies at Cambridge.

There were storms in the Arabian Sea that August, keeping everyone in their cabins, but when Chandra recovered, he had weeks of quiet cruising before him, several sheaves of paper, and a family habit of always using spare time productively. It was even an occasion when the usual racism of the Empire had its advantages: Chandra was a Brahmin with dark skin, and although the children of some of the

white passengers would try to play with him—
and he'd oblige—the parents would quickly
lead them away.

In the uninterrupted time at his deck chair,
he became one of the first to realize something very odd about the objects in the sky
above him. It was known that giant stars can
explode, with their top portions rebounding
away after they've collided with the heavy, collapsing core within. But what happens to that
remnant core, after the explosion?

Chandra was a cultivated young man, well
read in the literature of India and the West,
and especially fluent in German. He'd studied
Einstein's papers, and met a few of Germany's
leading physicists, on their trips to India. He
knew that the dense core of a star is under a lot
of pressure, and now he began to think about
the fact that pressure is a form of energy.

And energy is just another sort of mass.

Energy might be more diffuse than mass,
perhaps, but as $E=mc^2$ shows, they're both
just different versions of the same thing. Once
again, the two sides of the equation—the "E"
and the "m"—don't actually have to slip
across and "turn into" one another. Rather,
what the equation's really saying is that a
chunk of what we call mass actually is energy:
it's just that we're not used to recognizing it in
that guise. Similarly, a glowing or compressed

amount of energy really is mass: it just happens to be in a more diffuse form than we easily recognize as mass.

Chandra was about to glimpse the process leading to black holes. He just had to trace this logic forward as it spiraled in an escapable catch-22. A compressed star core is under a lot of new pressure, and that pressure is a sort of energy, and wherever there's a concentration of energy, the surrounding space and time will act just as if there's a concentration of mass. Gravity in the remnant star gets more intense, due to all this "mass." But that stronger gravity continues squashing what's left, so the pressure gets greater once more. Since the pressure can be treated as simply more energy, then—as Chandra now saw by the tremendous insight of $E=mc^2$—it acts as yet more mass. The gravity ratchets up.

In a small enough star, the buildup of pressure is low enough for the stiff material near the star's center to resist it. But if the star is massive enough, the process keeps on going. It doesn't matter how tough the star's material is; indeed, if it's exceptionally resistant, that will soon just make it worse. For suppose a giant star could hold up under even greater pressure than expected: immense, unthinkable, trillions upon trillions of tons bearing down. Well, that extra pressure would be more energy, which would mean it acts just as if it had more mass, and so the gravity would get even stronger, compressing it ever more.

Regardless of how hard the substance is at the core, the inside of the star will be crushed until . . .

Until what?

Chandra had all the openness to fresh thoughts of youth, but even he had to pause now. Could he be predicting that the inside of the star would actually disappear? If he was right, then rips were opening up in the very substance of the universe! He took time off for prayers and meals; he even spent hours politely listening to a Christian evangelist, who explained to this devout Hindu why all religions from India were the work of the devil. "He was a missionary," Chandra remembered later, "but he was also . . . anxious to please. Why be rude to him?"

When Chandra resumed work in his deck chair, he realized that he couldn't actually say what would happen to the remaining substance of the star, as it poured into the hole created by this never-ending collapse. But it was known in accord with other work of Einstein that space and time near the star would be strongly distorted by its presence. No light would ever escape; nearby stars that were pulled into its gravitational presence would get torn apart by what seemed an "empty" location in space.

This, along with other insights, was central to the modern concept of black holes. But once Chandra reached England, his vision was resisted by almost everyone he presented it to; often with less politeness than he'd granted the missionary. Eddington himself, the man who'd been so inspiring for Cecilia Payne, was now too old for any more such fancies. It was "stellar buffoonery," he declared. It was "absurd." But by the 1960s there was evidence of a star (look in the direction of the constellation Cygnus the Swan, and it's down a little to the left) that spins around an area that to our telescopes seems to be entirely empty space. The only thing that would be powerful enough to do this in so small a space would be a black hole. In the center of our own galaxy, there's even been evidence of another black hole, a truly monstrous one, which has accumulated to a great size over the aeons, swallowing, on average, the equivalent of one ordinary star each year. Space-time is actually being ripped open—as the young Chandrasekhar had been the first to see.

Chandra tried to fight Eddington's hostility in the 1930s, but when he found that even British astrophysicists who believed he was right were scared to back him in public, he ended up leaving England. He received a kinder welcome in America, and in an association with the University of Chicago went on to decades of work—culminating in his Nobel Prize in 1983, over a half century after that Arabian Sea voyage—which proved central in understanding what's in store for us next.

Five billion years from now, if Earth is flung loose from the fuel-emptied sun, any survivors or sensing devices left on our planet's surface will see a horizon darker than today's night sky. For the stars themselves will have used up their fuel and be dying out: the most fiery ones first, then the rest.

Earth's flight won't be stable through this darker expanse. Our Milky Way is already on track to collide with the Andromeda galaxy, and in several billion years, about the time of Earth's escape or immolation in the solar system, the great collision should finally happen. The spaces between stars are so great that most of the dimmed suns will just slowly pass through, without direct impact, but the turbulence will be great enough to shift an escaped Earth's trajectory once more.

If Earth slingshots inward, then in a few tens of millions of years we will be within range to be absorbed by the giant black hole at the galaxy's center. If we get slingshot outward, however, the end will simply be delayed.

By 10^{18} years from now (1 followed by eighteen zeroes, or 1,000,000,000,000,000,000 years from now), all galaxies will have emptied out because of such collisions. The black holes in the centers of the galaxies will slowly travel on their own, sucking mass and energy from the universe wherever they contact other objects. If it's another black hole that they randomly impact, then they simply merge, to become an even larger devourer. A few hours after coming within range of one of these, Earth and any distant descendants on it will be taken out of existence.

By 10^{32} years into the future, protons themselves will probably have decayed, and very little of ordinary matter will be left. The universe will be composed of just four classes of things. There will be electrons of the sort we're used to, with a negative electric charge, and there will be curious antimatter versions of electrons, with a positive charge, and there will be the swollen black holes, and even a cooled remnant of photons left over from the first seconds of creation, still traveling at their eternal 670 million mph speed after all these ages.

It doesn't end there, for, given enough time, even black holes can evaporate. Everything they engulfed will be released back—not in any recognizable form, but as an equivalent amount of radiation. At the very end, if the ordinary electrons and the antielectrons can cross the distances between them to annihilate each other, the last "mass" will have disappeared, leaving only "energy."

The universe will have returned in this way to a remnant of what it was at the start. For in the very first moments of creation, long before the sun was formed, the universe was dominated by the energy of radiation. Only later was there a net movement along $E=mc^2$, from the "E" side to the "m" side. The ordinary matter we're familiar with took shape out of pure energy, ultimately creating the stars and planets and life-forms we know. But now, near the end of time, over 10,000,000,000,000,000,000,000,000,000,000, 000,000,000,000,000,000,000,000,000,000, 000,000,000,000,000,000,000,000,000,000, 000,000,000 years into the future, it all goes back. Everything material that remains will travel from the "m" side to be on the original "E" side once again.

What will exist then is a dispersed radiation, spread over distances we cannot imagine, but the dominion of matter will be over. That was just an interlude in the final history of the universe. Now, mass and energy no longer transform into each other. There is a great stillness.

The work of Einstein's equation is done.

What Else Einstein Did

It wasn't actually $E=mc^2$ and his other work from 1905 that first made Einstein famous. If that were all he had done, his name would have become recognized within the specialized community of theoretical physicists, but probably not otherwise known to the public. In the 1930s he would have been just another distinguished refugee: living a quiet life perhaps, but in no special position to sign a letter warning of atomic dangers, which could be delivered to FDR in 1939.

It didn't turn out like that of course. Something else happened that built on $E=mc^2$ but went further—and ended up making him the most famous scientist in the world.

What Einstein published in 1905 only covered cases where objects are racing along smoothly, and gravity, with its accelerating pull, doesn't play much of a role. $E=mc^2$ is "true" in those cases, but will it hold true even if you get rid of those restrictions? That limitation and others had always troubled Einstein, and in 1907 he got the first hint of a wider solution: "I was sitting on a chair in my patent office in Bern when all of a sudden a thought occurred to me. . . . I was startled."

He later called this "the luckiest idea of my life," for a few years later, in 1910, it led to his reflecting on the very fabric of space, and how it was affected by the size or energy of objects at any one location in it. The work took several years, partly because although Einstein was in a league of his own in physics, he was only fair in mathematics. It wasn't quite as bad as he once described to a junior high school student in America, when he wrote her, "Do not worry about your difficulties in mathematics. I can assure you that mine are still greater." But it was enough to justify Hermann Minkowski's lament, when he saw the early drafts of Einstein's efforts: "Einstein's presentation of his subtle theory is mathematically cumbersome—I am allowed to say so because he learned his mathematics from me in Zurich."

To help him with the math, though, Einstein had his old friend from university days, Marcel Grossman, the one who'd loaned him crib sheets when they were undergraduates. (Grossman was also the friend whose father had written the letter getting Einstein the patent office job.) Grossman sat with Einstein for long hours, to explain what tools from recent mathematics he might use.

What Einstein's "lucky" idea of 1907 led him to, was that the more mass or energy there was at any one spot, the more that space and time would be curved tight around it.

It was a far more powerful theory than what he'd come up with before, for it encompasses so much more. The 1905 work had been labeled "special" relativity. This now was general relativity.

A small, rocky object, such as our planet, has only a little bit of mass and energy, and so only curves the fabric of space and time around it a bit. The more powerful Sun would tug the underlying fabric around it far more taut.

The equation that summarizes this has a great simplicity, curiously reminiscent of the simplicity of $E=mc^2$. In $E=mc^2$, there's an energy realm on one side, a mass realm on the other, and the bridge of the "=" sign linking them. $E=mc^2$ is, at heart, the assertion that Energy=Mass. In Einstein's new, wider theory, the points that are covered deal with the way that all of "energy-mass" in an area is associated with all of "space-time" nearby, or, symbolically, the way that Energy-mass = space-time. The "E" and the "m" of $E=mc^2$ are now just items to go on one side of this deeper equation.

The entire mass-loaded Earth rolls forward, automatically following the shortest path amidst the space-time "curves" that spread rippling around us. Gravity is no longer something that happens stretching across an inert space: rather, gravity is simply what we notice when we happen to be traveling within a particular configuration of space and time.

The problem, though, is that it seems preposterous! How can seemingly empty space and time be warped? Clearly that would have to occur, if this extended theory, which now embedded $E=mc^2$ in its wider context, were to be true. Einstein realized that there could be something of a test—some demonstration that would be so clear, so powerful, that no one could doubt that this wild result he'd come up with was right.

But what could that be? The proving test came from the heart of the theory, that diagram of a warp in the very fabric around us. If empty space really could be tugged and curved, then we'd be able to see distant starlight "mysteriously" swiveled around our sun. it would be like watching a bank shot in billiards suddenly take place, where a ball spins around a pocket and comes out with a changed direction. Only now it would occur in the sky overhead, where nobody had ever suspected a curved corner pocket to reside.

Normally we couldn't notice this light being bent by the sun, because it would apply only to starlight that skimmed very close to the outer edge of the sun. Most of the time the sun's glare would block out those adjacent daytime stars.

But during an eclipse?

$$E=mc^2 \cdot E=mc^2 \cdot E=mc^2$$

Every hero needs an assistant. Moses had Aaron. Jesus had his disciples.

Einstein, alas, got Freundlich.

Erwin Freundlich was a junior assistant at the Royal Prussian Observatory in Berlin. I wouldn't say he had the worst luck of any individual I've read about. Possibly there was someone who survived the *Titanic,* and then decided to try a ride on the *Hindenburg.* But it's probably pretty close. Freundlich was going to make his career, he decided, by shepherding the great general relativity equations forward, and performing the observations that would prove Professor Einstein's predictions were right. He was very generous about this—in the way that Lavoisier had been generous in letting his wife help him watch metal heat and rust. As a special honeymoon treat, Freundlich brought his new bride to Zurich in 1913 so that he could discuss stellar observations with the renowned professor.

An eclipse was predicted for the very next year, in the Crimea, and Freundlich prepared everything in detail. He even carefully arrived in the Crimea two months early, in July 1914. It was probably the worst possible place for a German national to be. War was declared one month later. Freundlich was arrested, put in prison in Odessa, and had all his equipment taken away. He finally got out in a prisoner exchange for a group of Russian officers who'd been arrested in Germany, but by then the eclipse had come and gone.

He didn't give up. In 1915, back in Berlin, Freundlich decided he could help Professor Einstein by measuring the way light got bent near distant binary stars. In February he had results that backed up the new theory, and Einstein began to spread the good news in letters to his friends. Four months later, though, Freundlich's colleagues at the observatory found he'd estimated the mass of the stars all wrong, and Einstein had to take it all back. For most people (as Freundlich's young wife perhaps tried to explain) that would have been enough, but Freundlich resolved to try again. Why didn't they try measuring how much distant starlight got deflected near the massive planet Jupiter; the one that the great Roemer himself had so persuasively used to resolve a scientific problem in an earlier era? Freundlich proposed it to Einstein. Einstein liked his earnest young helper, and in December he wrote to Freundlich's director at the Prussian observatory, suggesting that he be allowed to try this.

It would have been less painful just to have sent him back to the Crimean prison. Freundlich's superior was furious that anyone would dare to interfere. He threatened to fire Freundlich, insulted him in front of his colleagues, and made sure that he never, ever was allowed to get his hands on the equipment that could be used to test the prediction near the orbit of Jupiter.

But that didn't matter. Freundlich was hopeful again. A great new eclipse expedition was being planned, for 1919. If conditions allowed international travel, he'd finally be able to prove what he could do.

In November 1918 World War I ended. There were no obstacles to a German national traveling now! It's not recorded what Freundlich felt as the great expedition set out, but we know exactly where he was when the results came through. He read it in the newspaper, back in Berlin.

$E=mc^2$ · $E=mc^2$ · $E=mc^2$

In fact, it was a cool Englishman we've already met who led the team. Arthur Eddington wore small metal-rimmed glasses, was medium height and barely medium weight, and spoke in sentences that tapered off whenever he had to pause for thought, which was fairly often. This of course meant in the good English manner that under his meek exterior there beat a soul of wild determination. By the time Chandra encountered him in the 1930s his personality had hardened, but at this time, in the period of World War I, he had the energy of a young man.

On May 29 of each year the sun is positioned in front of an exceptionally dense group of bright stars—the Hyades cluster. That wouldn't usually help anyone, for without a solar eclipse occurring on that particular date, there would be no chance to see how that rich field of stars gets their light bent around the sun. The glare from the daytime sun would overwhelm that small effect. But in 1919 there was going to be an eclipse, precisely on May 29. As Eddington innocently

noted: "Attention was called to this remarkable opportunity by the Astronomer Royal
Frank Dyson in March 1917; and preparations
were begun. . . ."

What Eddington neglected to mention was
that he would have been thrown into prison if
he didn't go. For as a Quaker, Eddington was
a pacifist, and as a pacifist, in the middle of
World War I England, one of the rough prison
camps in the Midlands was in store. The soldiers guarding the pacifist camps were often
recently back from the front—or embarrassed
that they themselves hadn't seen service there,
which could be worse. Conditions were
rough. There was steady abuse and beatings; a
number of deaths.

Eddington's colleagues at Cambridge
didn't want him to go through this, and tried
to arrange for the War Department to defer
him, as being important for the nation's scientific future. A letter confirming this was sent to
him, from the Home Office, which he only
had to sign and send back.

Eddington knew what was in store in the
prison camps, but being a pacifist isn't the
same as being a coward, as the actions of many
Quakers years later in the American civil rights
movement showed. Eddington signed the letter, since that was only fair to his friends, but
then he also added a postscript, explaining to

the Home Office that if he wasn't deferred on grounds of scientific usefulness, he'd still ask to be deferred as a conscientious objector. The Home Office was not impressed, and began proceedings to send him to one of the prisons.

This is the point at which the Astronomer Royal, Frank Dyson, called attention to the remarkable eclipse opportunity. If Dyson could get Eddington to arrange the expedition, could Eddington still be deferred, despite that postscript? Dyson's work was relevant to navigation, and so he was close to the admiralty. The admiralty had a word with the Home Office. Eddington was free . . . so long as he led that expedition. They had two years to prepare.

It rained during the expedition, of course, but this is only what you'd expect on an island off the African coast, just north of the Congo, where Eddington ended up. But remember, Freundlich wasn't with Eddington. The rain cleared, and Eddington got two good plates. Most of the developing would have to be done in Britain, however, and no one would know the result for several months.

Afterward, Einstein tried to pretend that he hadn't been bothered by the delay. But by mid-September, still having no word, he wrote to his friend Ehrenfest, asking, with overelaborate casualness, if perhaps *he'd* heard anything about the expedition? Ehrenfest had good connections with the British. But no, he knew nothing. He wasn't even sure if Eddington had made it back.

In fact, Eddington had been back at Cambridge for several weeks now, but his photographic plates were a mess. They'd been carried by ship to West Africa, then kept in tents on a humid island, then carried into the rainstorm at the start of the eclipse, handled in and out of the camera, then brought back to the tents, and finally shipped by ocean steamer once more. The physical separations Eddington was looking for, in the movement of the distant stars, were going to be measured in tenths of a second of arc. On the small photographic plates, that came to fractions of a millimeter. (A thick pencil line is about one millimeter. If you have very good vision, you can just make out dust motes 1/20 of a millimeter across.) Eddington had micrometers to help, but Einstein would only be right if these tiny displacements were exactly as predicted, and so far, Eddington couldn't see them clearly enough to be sure if they were. The emulsion from the West Africa plates had become so jellylike in all the heat and transportation that if Eddington was honest he might never make out the necessary detail.

No one at Cambridge wanted to give up, though, for Einstein's was such a sweet theory. It was tremendous to think that the great tumbling ball of the sun was crashing down on the very fabric of space and time, sagging it so much that distant starlight started veering sideways as it go caught in the bend. Nor was it just the "traditional" mass of the sun that would be doing this. The 1905 equation entered in also. All the heat and radiation blasting out of the sun—all that "energy"—was acting as an additional form of "mass." It added to the bulk of the sun as well. (This was at the heart of what Chandra would build on, in his later sea voyage of 1930.)

Luckily, the British Empire had its traditions, and one of the prime ones was that something always went wrong. Explorers, conquerors, younger sons and even metal-eyeglassed Quaker astronomers had learned that lesson: picking it up from a lifetime of hearing about one imperial expedition after another.

And that's why Eddington had sent out a second team—an entire duplicate crew—to be sure he proved Einstein's prediction.

This second crew had a different telescope, and had been sent to a different continent (they'd been in northern Brazil), and they even had a different mechanical drive for the telescope. It was all in the finest tradition of spreading the odds, and it worked. Once the Brazil team's plates came back, and a special oversized micrometer had been built to fit around their larger plates, and Eddington and the others had measured and remeasured, the congratulatory telegrams started bursting out. Bernard Russell, who had recently been a Fellow at Trinity, now received a message from his old friend Littlewood: "Dear Russell: Einstein's theory is completely confirmed. The predicted displacement was 1".72 and the observed .1"75+/− .06."

E=mc² · E=mc² · E=mc²

The celebration was in style. The Royal Astronomical Society was invited to a joint session with the Royal Society on November 6, 1919, in the great room at Burlington House, on Piccadilly. Scientists came in from Cambridge to the stations at King's Cross and Liverpool Street; cabs were taken; nonscientists who'd heard something momentous was to be announced arrived as well. A visitor described the evening: "There was dramatic quality in the very staging:—the traditional ceremonial, and in the background the picture of Newton to remind us that the greatest of scientific generalizations was now, after more than two centuries, to receive its first modifications."

Dyson spoke, and Eddington spoke—there's no record if any narrow-eyed parole officer from the Home Office was in the room—and then the elderly chairman stood up to speak:

> This is the most important result obtained in connection with the theory of gravitation since Newton's day, and it is fitting that it should be announced at a meeting of the Society so closely connected with him....
>
> If it is sustained that Einstein's reasoning holds good ... then it is the result of one of the highest achievements of human thought.

With World War I just over, these findings were wondrous. God may have seemed lost after the trenches, but now order had been divined in the cosmos. Even better, a German and an Englishman working in harmony had found it. Royalty and generals and political leaders and even artistic figures who'd made their reputation under the old regime—the regime that had led to the slaughters of World War I—were discredited. The categories of "people to respect" were nearly empty. Einstein, instantly, was the greatest media celebrity on the planet. Headlines in *The New York Times* for November 10, 1919, announced:

> "Light All Askew in the Heavens: Men of Science More or Less Agog Over Results of Eclipse Observations."

and

> "Einstein Theory Triumphs: Stars Not Where They Seemed or Were Calculated to Be, but Nobody Need Worry."

This meeting was also when the rumors began that only a dozen people could understand what it all meant. The *New York Times* did have a few knowledgeable science writers, but they were in New York. The London bureau was handed the story, and Henry Crouch was asked to cover Burlington House. In the history of inappropriate assignments this is at the Lyman Briggs level. Crouch was a good journalist in the sense that he knew you had to make a story interesting. He was somewhat less good, however, in having the slightest clue what was going on here—Crouch was the paper's golfing specialist.

But he was also a *Times* man through and through, and nothing like a simple lack of knowledge was going to hold him back. He kept on filing, and the headline writers pulled out the key parts of his story:

> "A Book for 12 Wise Men: No More in All the World Could Comprehend It, Said Einstein When His Daring Publishers Accepted It."

He made that up. Einstein was writing no book, there were no publishers involved—daring or not—and most of the physicists and astronomers attending understood easily enough what the meeting was about. Crouch had started the theory off on its track record of poor public comprehension, from which it never entirely recovered.

And that only added to the fame. In almost all religions, there's a powerful difference between a priest and a prophet. A priest merely stands below an open hole in the sky, and lets the truth that's normally kept hidden up there come pouring down. (Press secretaries and nuclear technicians are examples.) A prophet, however, is someone who manages to journey up through that opening. They are individuals who can venture to that Other Side, before returning back to ordinary life, here with us on Earth. As a result, we'll try to glimpse, in the expression on their face, or in the simple equations they've plucked and brought back down, what things are like up there, in that higher realm, which so many of us believe in, but know we'll never get to visit directly.

Martin Luther King Jr. and Nelson Mandela have been considered such prophets, carrying down a vision of racial harmony; their words spreading afterward with a power that came from their being felt to have originated in that higher source. In post-World War I Europe, Einstein's findings were received with the veneration King's or Mandela's words would be granted later. And since very few people understood Einstein's work at first, all the feeling it suggested—all the desire for transcendence and for knowledge from Einstein's mist shrouded divine library—would soon be shifted onto images of Einstein himself. Perhaps that's why people were attracted to photos of him that had a distinctive, sadly bemused look. They matched the later most powerful photos of Martin Luther King, where he too seemed to be sadly seeing something greater than ordinary mortals could.

Einstein tried to push back some of the fame. He called the exaggerating newspaper accounts "an amusing fear of imagination." Two weeks after the public announcement, he wrote in the London *Times* that although the Germans were proudly calling him a German, and the English were proclaiming him a Swiss Jew, if his prediction ever came to be shown false, the Germans would call him a Swiss Jew, and the English would call him a German. In fact, he got it wrong: his astronomical prediction and the 1905 equation both stayed true, but English anti-Semites such as Keynes still scorned him ("this ape-like Jew, attempting to appear civilized"), and with the rise of Hitler the German government not only called him a Jew but supported the calls to have him killed. After leaving the Continent, and trying England, he ended up in America for the rest of his life: in 1939 signing the letter to President Roosevelt, which, albeit indirectly, helped lead to the atomic bomb; otherwise just living a quiet professorial life at 235 Mercer Street, in Princeton, New Jersey.

He never especially liked the Ivy League snobbery of Princeton ("this village of puny demi-gods upon stilts," as he described it to a European friend). There were giggling bobby-soxers; the occasional gaping tourist; at the Institute for Advanced Studies—a two-mile walk from his home, which he took regularly—younger scientists kept a surface politeness, but he knew that many disparaged him behind his back as someone too old to be useful.

That alone didn't seem to bother Einstein. His goal, as always, was simply to see what had been intended for our universe by The Old One. What he had scribbled in his now-yellowing manuscripts decades before, as well as the new equations he was constantly working on now—trying to create a theory that would unify in a clear and predictable way all the known forces in the universe—still seemed, to him, the best possible track forward.

What did hurt him were different reminders of how things had worked out. One, almost too horrible to think of vividly, was implicitly brought up each time he encountered Oppenheimer, his institute's head, who'd led the Manhattan Project that had demonstrated that $E=mc^2$—despite Einstein's lack of involvement—could be turned into vast fields of death in Hiroshima and Nagasaki. "Had I known that the Germans would not succeed in producing an atomic bomb," Einstein once told his longtime secretary, "I would never have lifted a finger. Not a single finger!"

Then, as the years went on, there was the increasing feeling of his own powers fading away. An unintentionally tactless young assistant once asked him about this. Einstein explained that it was more difficult to judge which of his ideas were worth pursuing—a great contrast with his

younger years, when he'd been superb at identifying the key issues in a field. "Discovery in
the grand manner is for young people," he'd
once told a friend, ". . . and hence for me a
thing of the past."

He settled into an old man's daily routine,
in his simple white suburban house on Mercer
Street. His sister, Maja, was with him in
America by now. She suffered a severe stroke
in 1946, and virtually every evening after that,
for the six years till her death, Einstein would
drop whatever work he was doing, and climb
the stairs to her room, where he would read
aloud to her for hours. Before that, most days,
there were the mock-chiding rituals with his
housekeeper; the saddened disregarding of reminders of his mentally disturbed second son;
sometimes the memory of the afternoon when
he'd played one of the violins in Mozart's G
minor Quintet with those nice youngsters
from the Juillard Quartet. But there were also
the moments, settled comfortably in his upstairs study, when his steadily penciled pages of
symbols lifted him back into his past; to the
time when anything had seemed possible.

And the works—of the divine library that
he was convinced awaited—could once again
be read.

APPENDIX

Follow-up of Other Key Participants

When MICHAEL FARADAY took over Davy's position as director of the Royal Institution, he moved permanently into the Royal Institution with his wife. He continued to make major discoveries well into his fifties, but despite many requests, he never took on a personal student.

After the execution of ANTOINE-LAURENT LAVOISIER, his remains were carted out of Paris, passing through one of his new tollgates that had survived the 1789 attacks. A few months after Lavoisier's death, the body of the man who'd ordered the executions, Marat's colleague Robespierre, was carted through the same gate, and placed in the same common grave. It was a converted wasteland called "Errancis" ("maimed person"), which they now shared. Several fragments of the solid tollgates from the GENERAL FARM wall, which Lavoisier had ordered built, can be seen to this day, in the Parc Monceau, and near the metro exit at Denfert-Rochereau.

A few months before Lavoisier's arrest, a young woman, Charlotte Corday, called at the apartment of JEAN-PAUL MARAT, asking to see him. His guards refused, but when she insisted she had news about dangerous political opponents, he overruled them, and had her let upstairs. Since Marat had skin complaints that forced him to spend much of the day in a bathtub, it was from that position that he greeted her, discovered that the political opponents were members of her family (whom he had ordered killed), and then saw her step forward, knife out. She stabbed him to death, in an assassination later immortalized by the society painter David.

Since MARIE ANNE PAULZE had been only thirteen when she married Lavoisier, she was just thirty-five when her husband was killed. Although harassed by the Revolutionary government, and her wealthy apartment emptied, she outlived most of her persecutors, and enjoyed a peaceful old age.

When he went back to Denmark, OLE ROEMER quit all astronomical research, though he was only in his twenties. He married the daughter of his ethics professor—the man who'd first brought him to the attention of Cassini's scout—and ended up becoming chief constable of Copenhagen, then director of the port, and for several years the equivalent of a Supreme Court judge. In his spare time he worked on an improved device for the measurement of temperatures, which a visiting businessman named Daniel Fahrenheit thought had some merit. Roemer died in 1710, seventeen years before the British experiments that finally proved he'd been right about the speed of light.

JEAN-DOMINIQUE CASSINI outlived Roemer, and continued to promote only those astronomers who—erroneously—agreed with him that light traveled at an unmeasurable speed. The dynasty he established ran for almost two centuries, up till the fourth generation, ending with the Cassini who was forced to close down his great-grandfather's proud Observatory — the building Lavoisier had seen from his prison window.

In 1997 the European Space Agency (ESA) launched a probe, that in late 2000 would reach the orbit of the planet Jupiter, which

Roemer had used for his epochal prediction. The probe was proudly named *Cassini*. France is a major funder of the ESA.

VOLTAIRE lived to extreme old age, writing and mocking all the way. His collected works run to over 10,000 printed pages, and did much to promote the Revolution, which began just a few years after his death. He never published significant commentaries on science after du Chatelet's death.

The manuscript that EMILIE DU CHÂTELET finished in her final days—*Principes Mathématiques de la Philosophie Naturelle*—became a great success in the scientific circles of its time. A first edition can be viewed at the Bibliotheque Nationale in Paris. The Chateau de Cirey ended up shuttered and abandoned during the Revolution, but was later refurbished. Her first son never lived to see that, having become ambassador to Britain under Louis XVI, which led, after his return to France, to his arrest and subsequent death by the guillotine. "If I were King," du Chatelet once wrote, ". . . women would be worth more, and men would gain something new to emulate."

HENRI POINCARÉ lived for seven years after Einstein's 1905 publications, still unreconciled to the fact that outside of France he wasn't recognized as a founder of relativity. In his final years he wrote eloquent, thoughtful essays on creativity. He also ensured that no one who wanted to work on Einstein's theories could be promoted in France.

MILEVA MARIĆ-EINSTEIN continued looking up to her husband, even as he started an affair and their marriage broke apart. When he promised to give her any future Nobel Prize money as a divorce settlement though, she saw nothing unusual in assuming that he would win it. (In 1922, when he did get the prize—though not for his theory of relativity, as the Swedish Academy was still not entirely convinced it would prove itself—he promptly transferred to her the substantial prize money as promised.)

She never remarried after the divorce, and having missed her chance to retake her final university exams (her grades had been just slightly too low to get a teaching job), she never found a significant career. Although her first son became an engineering professor at Berkeley, she became exhausted caring for their second son, who was in and out of mental institutions his whole life. She died in Zurich in 1948, increasingly depressed and alone.

MICHELE BESSO, Einstein's closest friend from his Bern years, with whom the ideas of special relativity were first talked over, had a rich home life, and a successful career as a mechanical engineer. In the 1950s, when he and the now twice-married Einstein were old, they began to write each other more frequently. After Besso's death, in early 1955, Einstein wrote Besso's family: "The gift of leading a harmonious life is rarely joined to such a keen intelligence, especially to the degree one found in him. . . . what I admired most about Michele was the fact that he was able to live so many years with one woman, not only in peace but also in constant unity, something I have lamentably failed at twice. . . ."

Despite the bowling balls and child's hoe that had been thrown her way as a child, MAJA EINSTEIN became her big brother's closest friend. In 1906 she moved to Bern in part to be close to him, and ended up taking a doctorate (in Romance languages) from the university there, an extremely rare achievement for a woman at the time. When Einstein began teaching at that university she—and Besso—regularly attended some of his first classes, so that the authorities would be less likely to notice how few other students Einstein was then getting.

ERNEST RUTHERFORD died suddenly in 1937, following an intestinal rupture, which was possibly linked to overvigorous gardening he was doing at his weekend cottage. His final words were for his wife to be sure to arrange that scholarship funds be sent to Nelson college, in New Zealand, which is where he had received the schooling that raised him from rural poverty, and prepared him for his own scholarship to England. THE CAVENDISH LABORATORY he left behind never again achieved the same preeminence in nuclear research. In time a new director shifted it increasingly toward biology. This included welcoming a young American, James Watson, who it was thought might work well with the physics-trained Francis Crick, in using the Cavendish's resources to try investigating the structure of DNA.

HANS GEIGER, Rutherford's young man who'd had the knack of making such useful radiation counters, returned to Germany, and soon assumed senior academic positions. His years in England had, however, little effect in making him a believer in tolerance of freedom. He was one of the most active of senior German physicists in supporting the rise of Hitler, and welcoming students with swastikas. He turned against his Jewish colleagues, including ones who'd helped him over the years; as Hans Bethe and others have noted, he seemed to

enjoy coldly turning down any of their requests for aid in obtaining foreign posts.

SIR JAMES CHADWICK was holidaying with his
family on the continent when the German invasion of Poland began in 1939, and although
he was assured by his hosts that there was no
chance of being caught behind enemy lines,
he brought his family back to England with
remarkable alacrity. Having stood up to
Oppenheimer enough to impress General
Groves, he was brought into the centers of
power in Washington, and turned out to be
one of the most effective administrators of the
Manhattan Project. He lived into the 1970s,
but was so distressed by what the bomb explosions could lead to that "I had then to start
taking sleeping pills. It was the only remedy.
I've never stopped since then. It's 28 years,
and I don't think I've missed a single night in
all those 28 years."

ENRICO FERMI had related well to virtually everyone he worked with in Italy, and repeated
the process in America. He worked hard to
master American colloquialisms, and admitted
failure in his Americanization efforts only
when it came to clearing the lawn of his first
suburban house of crabgrass—for did it not,
he and his wife inquired, have as much of a
right to grow there as anything else?

His participation in the Manhattan Project was central to its success, but as with a number of the participating scientists, he was hit by cancer when he was still only in middle age. He was notably calm in the hospital room during his last few months. When the Hindu Chandrasekhar came in, unsure what to say, Fermi put him at ease by asking, with a smile, if Chandra could tell him if he was going to come back as an elephant the next time.

America's largest high-energy physics research center is located about 100 miles southwest of Chicago. It is called FERMILAB.

OTTO HAHN received the Nobel Prize for the work that Lise Meitner had led him toward. Instead of explaining that this was a mistake, and that she should have been honored, even if only jointly, he began to write her out of the story. In his first postwar interviews he started saying that she had merely been a junior research assistant; later he pretended (believed?) that he'd barely heard of her at all.

For many years, the workbench Meitner had used in Berlin, with all the devices she'd accumulated for the key experiment, was on display in the Deutsches Museum in Munich. It was labeled the *Arbeitstisch von Otto Hahn*: The workbench of Otto Hahn.

As a mark of Hahn's fame, when a new chemical element, number 108 in the periodic table was created, it was given the provisional name Hahnium.

FRITZ STRASSMAN was disappointed at Hahn's antics, and refused the 10 percent of the Nobel Prize money that Hahn later offered him. He kept his liberal sympathies even in the midst of the war, hiding the Jewish pianist Andrea Wolffenstein for several months in his Berlin apartment—for which he was later honored at the Holocaust memorial Yad Vashem, in Jerusalem. After the war Strassman wrote to Meitner, asking her to return to Germany, but noting that he'd understand if she didn't.

LISE MEITNER was hurt at what her lifelong partner Hahn did to her, but put it down to his desire to suppress everything about the recent German past. She left Stockholm for Cambridge, England, and in the 1960s could be seen as a slender very old woman, browsing in the bookshops. Into her mid-eighties she kept a notebook of questions to ask her young nephew. These included topics in current theoretical physics, as well as perplexing vocabulary words such as *highfalutin'* and *juke box*. She died in relative obscurity in October 1968, a few weeks after the world-famous Hahn.

In the 1970s feminist scholars began to reexplore her career. When a new chemical element, number 109 in the Periodic Table, was created in 1982, it was named Meitnerium.

The young nephew, ROBERT FRISCH, managed to get out of Denmark before the German army invaded. Successfully reaching England, he was barred from classified work on radar because he was an enemy alien, and so had time for the computations that showed that much less uranium than suspected would be enough for a bomb. This was the basis for the classified report that jump-started the U.S. bomb project when it was finally brought out of Lyman Briggs's safe.

Frisch played an important role at Los Alamos, though by March 1945 he was back in Cambridge, where he happened to be at the Cavendish laboratory when the young Fred Hoyle came by, in need of some listings of nuclear masses for an idea he'd had about the way elements were formed inside stars. Frisch supplied them.

After the war, with his first name now "Otto," Frisch became a firm anglophile, though he always retained a suspicion that "the weather" was something only newly arrived in Britain, this being the only reasonable explanation as to why the populace commented on it so frequently. To his great pleasure, in 1947 he was offered a professorship at Cambridge—so allowing him to share in the tradition that an earlier immigrant, Ernest Rutherford, had begun.

As soon as the bombs to be used against Japan were delivered, J. ROBERT OPPENHEIMER went back to being as sarcastic as ever, suddenly addressing the staff who remained at Los Alamos as "second raters." He also applied his sharp tongue to Lewis Strauss, head of the new Atomic Energy Commission (AEC), as well as to Edward Teller, which meant that he had serious enemies when a witch-hunting AEC committee investigated his 1930s attendance at left-wing parties, as well as his moral reticence about the hydrogen bomb. In 1954 he was purged from all government service.

LESLIE GROVES always kept a soft spot for Oppenheimer. Retired from the army, and an executive at Remington Rand, he refused to condemn Oppenheimer wholeheartedly (as most other army staff did) at the 1954 hearings. Groves always held that Oppenheimer was "A real genius. . . . Lawrence is very bright, but he's not a genius, just a good hard worker. Why, Oppenheimer knows about everything. He can talk to you about anything you bring up. Well, not exactly. I guess there are a few things he doesn't know about. He doesn't know about sports."

Using material from Lawrence's lab, EMILIO SEGRÈ had become the first person to create the element technetium. He also managed to stay at the Berkeley Lab long enough to become the codiscoverer of plutonium, the element used in the Nagasaki explosion. At the reduced salary Lawrence gave him, there had been no chance of bribing any consular officials to get his elderly parents out of Italy. His mother was captured during a Nazi manhunt in October 1943, and murdered soon after that; his father, who had been safely hidden in a papal palace, died of natural causes the next year.

When the war was over Segre went to his father's tomb, scattering a small sample of technetium from Lawrence's lab over it: "The radioactivity was minuscule, but its half-life of hundreds of thousands of years will last longer than any other monument I could offer."

As soon as Denmark was liberated, GEORGE DE HEVESY went back to the jar of strong acid in which he'd dissolved the Nobel gold medals at Niels Bohr's Copenhagen institute, and simply precipitated them back out. The Nobel foundation then recast them, and they were returned to their rightful owners. When de Hevesy had first dissolved them he'd only just recovered from a full-fledged midlife crisis, convinced that at age fifty he was past the age for fresh invention. The recovery was quite complete, for soon he had a Nobel medal of his own, awarded for work he did—at an age when most physicists' creativity is long gone—on radioactive tracers.

All laureates are offered Swedish citizenship, but de Hevesy was one of the few who took that up, settling in Stockholm for the rest of his long life. In the 1960s, he could sometimes be seen strolling in La Jolla, California, an erect elderly man, happy visiting with his American grandchildren, telling them what he remembered of life growing up in the 1880s in a baronial palace in Hungary.

ERNEST LAWRENCE came out of the war in triumph, and succeeded in raising more and more funds, and building larger and larger machines, until finally he proposed a cyclotron that violated the special theory of relativity, and so was physically impossible. None of his young men would dare to explain that to him, however, and the failure of his efforts to get it to work ended up wrecking his health. A little before he died, in 1958, he told a group of graduate students at the University of Illinois: "Why, fellows, you don't want a big machine. There's too much emphasis these days on sheer size for its own sake."

WERNER HEISENBERG became the grand old man of German science, and after a brief six-month internment in the luxury of a grand country house in Cambridgeshire, England, was soon respected worldwide as a sage and philosopher. He rarely spoke of the war, but when he did, would give the impression through hints and nodding gestures that he had been able to make a bomb all along, but had willfully led the research in the wrong direction, to keep the Nazi government from getting the weapon.

Heisenberg never realized he was being recorded at the Cambridgeshire country house:

HEISENBERG: "Microphones installed? laughing Oh no, they're not as cute as all that. I don't think they know the real Gestapo methods; they're a bit old-fashioned in that respect.

But when the recordings were released a half century later they proved Heisenberg's cover story false. There was a fine justice in Heisenberg's and the others' being sequestered there, for it was only a short distance from the other elegant country house that the British secret service kept, where the six Norwegians who destroyed his project had prepared for their mission.

Heisenberg almost hadn't survived to be captured, for the predecessor of the CIA had sent an assassin, the ex-athlete Moe Berg, against him during that final Swiss trip. Berg was planted in the audience of the seminar Heisenberg gave in Zurich. If Heisenberg showed evidence that his bomb project was on the right tracks, he would be killed. Berg had a gun, and understood some undergraduate-level physics, but the talk was too technical for him to follow. His scrawled notes from the meeting that survive in CIA archives: "As I listen, I am uncertain—see: Heisenberg's uncertainty principle—what to do to H. . . ." He left Heisenberg alone.

KNUT HAUKELID survived the war, despite the vast manhunt that began after his sinking of the Lake Tinnsjo ferry. Transcripts from Heisenberg's internment finally clarified the significance of that sinking, where the equivalent of about 600 liters of concentrated heavy water had gone down. (In the following extract, Heisenberg is speaking in English, thus the imperfect grammar):

HEISENBERG: We have tried to make a machine which can be made out of ordinary uranium....

(QUESTIONER): With a little bit of enrichment?

HEISENBERG: Yes. That worked out very nicely and so we were interested in it.

(Pause)

After our last experiments, if we had 500 liters more heavy water, I don't doubt that we had got the machine going....

Haukelid became an officer in the Norwegian army; another member of the original commando team (Thor Heyerdahl at ease by sailing with him on *Kon-Tiki*.

The heavy water facility at VEMORK continued in operation till the early 1970s, when, having outlived its economic usefulness, it was blown up by Norsk Hydro engineers. Some of the rubble was removed by truck and train, but much was left in place, and simply paved over. Several thousand visitors walk over it each year, for the old generating station behind it has been converted into an excellent museum, and the location of the commando raid is directly under the route to the entrance.

I. G. FARBEN company, which had taken over the plant's operation during the war, was briefly broken up by Allied authorities, after the Nuremberg trials showed its executives profiting from the purchase and subsequent death of human slaves. One of its main constituents, the BAYER company, though popularly known just for its aspirin, continued to be a major force in general chemicals worldwide.

The BERLIN AUER factories, where female slaves from Sachsenhausen had been worked to death to supply Heisenberg with uranium oxides, almost survived intact till the end of the war. In the last few months, though, they were obliterated by Allied bombers acting on Groves's instructions, in large part to keep them from falling into Russian hands. Almost all the Berlin Auer executives avoided jail sentences, and indeed even before the war ended had been thinking of their future. American investigators found that all Europe's supplies of radioactive thorium had been purchased by an unknown buyer—it was the Berlin Auer company, which planned to use it to make white-glowing toothpaste once more.

War crime trials in Oslo after the war led to the conviction and imprisonment of several of the guards—both Germans, and Norwegian collaborators—responsible for the deaths of the surrendered BRITISH AIRBORNE TROOPS. Many of the troops had been thrown into shallow graves, with their hands tied behind their backs with barbed wire. When they were disinterred for reburial in Britain, the head of the Norwegian collaborationists government, Vidkun Quisling, was forced to help dig up their remains.

The once-secret reactor at HANFORD, WASHINGTON, which had played such an important role in creating the plutonium used in the Nagasaki and later bombs, continued as a central site for the production of American nuclear weapons. After several decades of service, though, a changed national mood increasingly saw it as a center of environmental despoliation: cleanup costs for its leaked or inadequately stored radiation were estimated at $30-$50 billion.

CECILIA PAYNE'S thesis advisor brought her career to a halt by making sure she was kept from any of the new electronic equipment coming in. He also ensured, as director of Harvard's observatory, that when she did give courses, they weren't listed in the Harvard or Radcliffe catalog; she even found out, years later, that she had been classified as "equipment expenses" when her salary came due. When the worst of the sexism ended, and a decent director of the observatory took over in the post-war era, it was too late. She had such a heavy teaching load by then that "there was literally no time for research, a setback from which I have never fully recovered."

Instead, she became one of the kindest supporters of the next generation at Radcliffe, always available for long talks to students at loose ends. She also kept intellectually nimble by learning languages, to add to the Latin, Greek, German, French, and Italian she'd been comfortable with when she'd arrived in America. "Icelandic was a minor challenge," her daughter wrote, though "I cannot say she truly mastered it." Cecilia Payne had the pleasure of seeing that daughter become an astronomer—and publishing several papers with her.

ARTHUR STANLEY EDDINGTON became increasingly resistant to the main trends of modern astronomy. One of his final works, published in 1939, had a chapter beginning "I believe there are 15 747 724 136 275 002 577 605 653 061 181 555 468 044 717 914 527 116 709 366 231 425 076 185 631 031 296 protons in the universe, and the same number of electrons."

He was perplexed that professional astronomers stopped paying any attention to him.

In 1950, four years after FRED HOYLE'S paper on bomblike implosion inside stars, the merits of Cambridge nepotism were demonstrated when a director of radio talks from his old college overlooked the stern injunction against Hoyle in BBC files, and invited him anyway to give a series of broadcasts on astronomy. In the rush to prepare a script for the final talk, Hoyle coined a somewhat mocking phrase for a then-unproven theory about the origins of the universe. He called it the "Big Bang."

The BBC talks and subsequent book were such a success that not only did Hoyle and his wife get enough money to buy their first refrigerator, but it led to a career popularizing science, which he carried on in parallel with his academic research. This allowed him to put enough savings aside that in 1972, when he told Cambridge administrators he would resign if they continued going back on their word about funding for the successful astronomical research center he'd created, he was able to startle them ("Fred won't resign. Nobody resigns a Cambridge Chair"), and politely walk out. He has continued to publish innovative papers, some of them flighty, some of them profoundly sensible—as has been the wont of top scientists from Newton on. If it weren't for the way his Yorkshire honesty irritated the old guard in Britain and the astronomical community generally, it's generally accepted that he would have long since been granted a Nobel Prize for his work on the formation of the elements.

SUBRAHMANYAN CHANDRASEKHAR was renowned for keeping a calm exterior, but internally: "I am almost ashamed to confess it. Years run apace, but nothing done! I wish I had been more concentrated, directed and disciplined." At the time of this lament he was twenty, and it was but one year since the sea journey where he'd peered into the catch-22 from $E=mc^2$, which, along with other work, would ultimately lead to the understanding of black holes. He accepted a post at the University of Chicago, but his reserve meant that he and his wife settled in an observatory town 140 miles from the main campus, largely so that they wouldn't have to embarrass Chicago faculty members by turning down invitations where alcohol or meat might be served. He diligently drove the full round-trip journey to Chicago for his teaching when needed, even during winter storms

—once for a class that had only two students. (It was worth the drive, as that entire class— Yang and Lee—went on to win the Nobel Prize.)

Forty years after his rebuff by Eddington, Chandra finally returned to the study of black holes. There were photographs of brightly dressed young physicists in the clothes of the early 1970s, sitting around a table in the Caltech cafeteria, listening to this perfectly tailored, suited man of the generation of their grandparents. He surpassed almost all of them in his agility with new applications for general relativity, and in 1983, over half a century after the sea voyage, he published one of the fundamental works on the mathematical foundations of black holes. That was the year he won the Nobel Prize, and then—following his usual habit—he shifted directions once again, beginning an elegant exploration of Shakespeare, and of esthetics generally.

In mid-1999, NASA launched a large satellite for deep space observation, capable of capturing images from the very edge of black holes. The satellite crosses over much of the earth— the Arabian Sea, Cambridge, and Chicago included—in its mission, and it is called the Chandra X-ray Observatory.

Although ERWIN FREUNDLICH missed out on the 1919 eclipse expedition, his spirits recovered when industrialists in the new Weimar Republic donated large funds to build a great astronomical tower in Potsdam. This would allow tests of general relativity's predictions, even in periods when there was no eclipse. Zeiss supplied the equipment, and Mendelsohn, the great expressionist architect, designed the building—it's the famous Einstein Tower featured in many books on 1920s German architecture.

Through Einstein's help, Freundlich became the Einstein Tower's scientific director. The measurements he undertook, however, proved to be impossible with the technology of the time. Only in 1960, at Harvard did another team manage to give this further confirmation of Einstein's work.

Acknowledgments

I couldn't have written this book on my own. A lot of it developed out of the Intellectual Tool-Kit courses I taught at Oxford, which Roger Owen and Ralf Dahrendorf were central to getting started. Avi Shlaim helped nurture that series over the years, and Paul Klemperer made apt comments after one of the creativity lectures, which helped lay the idea for some sort of expansion of the physics aspects of that course.

Once a first draft was done, several friends were kind enough to read the manuscript in its entirety: Betty Sue Flowers, Jonathan Rowson, Matt Hoffman, Tara Lemmey, Eric Grunwald, Peter Kramer, and Caroline Underwood. They gave excellent suggestions, a number of which I even adopted.

George Gibson and Jackie Johnson at Walker & Company were even more valiant: repeatedly offering wise comments that improved the book no end. Readers who looked through particular chapters for accuracy included Steven Shapin, Dan ver der Vat, Shaun Jones, and Frank James. None of them, of course, are responsible for any errors that remain.

Two individuals gave especially important structural comments. In a series of long, flowing phone talks Doug Borden helped me see how the final visions of the "energy" and "mass" chapters could be best developed. Gabrielle Walker, the most eloquent of friends, talked me through all aspects of the book, opening up a world of honest writing. In one particularly memorable late-night stroll through St. James's Park, she explained how the quietly widening chorale of the St. Matthew's Passion showed the way to escape from strict chronology after the equation's story reached 1945. The book would have collapsed after chapter thirteen without that.

For a long time I was perplexed about what level of explanation would be best in the main chapters. Peter Kramer, especially, was persuasive in his observation that I needed to give the results of the equations, without slighting the explanation of why the equation holds true. To do this, I put an indispensable core of explanation in the main text, a little more in the notes at the end, then even more—and especially anything that involves mathematics—is in the website TK. I like the idea that a book is no longer a single defined object, limited by the technology of paper and glue and stitching. To keep the website from being only for technical types, I also included there some reminiscences of boyhood in Chicago (which with only a slight twist lead to an explanation of how space and time slosh into each other. There are also insights from William Blake, samples of Einstein's voice, links to the courses I offer on the equation and other matters, a look at why simple art forms such as equations are so often true, and other odds and ends.

The newly finished British Library was an excellent place to research all this: It's one of the great libraries of the world, and possibly the last, pyramid-like homage to the pre-Internet era. Many of the Library's science journals were in the old Southampton Row reading rooms, where interior design and coffee facilities are not quite at the same level, but the photostats of original patent applications on the wall (Whittle's jet engine, the paperclip, the thermos flask, the Wright brothers' wing-warping) made up for a lot of that.

The University College science library in London was also useful, and even though the physical plant is now showing the effects of years of underfunding, the staff do an excellent job of trying to shore up the gaps. The London Library on St. James's Square doesn't suffer those funding problems and is a strong reason for living in this city. It's an early Victorian institution that still works: There are about a million books, on open-shelves, including many early editions. I became used to reading texts that would refer to some earlier biographer's hard to obtain work, which could usually be conveniently found, albeit under a l ight sprinkling of dust, just an arm's length further along the shelf.

There was an added benefit, since for Faraday and Maxwell and the like I could scoop up armfuls of their works or letters, and head outside to one of the benches under the oaks in the center of St. James's Square. It was a fitting location. To one side was the red-brick building which had housed Eisenhower's SHAEF headquarters in 1944, when the fears of a German atomic bomb were near their peak; behind me was the plaque to Ada, Countess Lovelace, the 19th century predecessor of computer programmers, who experienced many of the ups and downs a woman's career in science was likely to take.

Most of the actual writing was done when my wife Karen was making a transition from being a distinguished historian, to being a distinguished business consultant. We'd always spent a lot of time with our children, but when she was off in Geneva or Washington or Berlin—although she later helped with draft after draft—I had even more of the day with them. This meant writing time was often broken up. But curiously the text proceeded faster than before. What happened, I think, was that by really getting into the time with the kids, I was forced to have the break times that authors rarely allow themselves. Strolling to school we'd get down on our bellies to observe ants in the grass, or we'd stop and chat with the men drilling the streets, who almost always had younger brothers and sisters, or kids of their own, and so were only too happy to rest and explain how their tools worked to the fascinated three and five year olds. There'd also be wall walking and "secret spy," long lunchtimes and afternoons. There were times when I was grumpily distracted (sorry guys), but most of the time I looked forward to our hours together, and the tender refreshment that very young, very curious minds provide (thanks guys).

And when it finally did get too late for more, and two exhausted youngsters were asleep in their bunk beds, I'd settle into a big chair in their room (it felt a lot friendlier there than being in my study), with notes and bound volumes spread out, and then I'd gladly return for hour after hour to this book, as the sky darkened and the London streets went quiet outside. A few times—the writing racing along; my coffee long since cold—I'd realize I'd gone the whole night through; most notably once while writing about the chemistry of the sun, as the roaring sphere of that star—powered by thermonuclear blasts in accord with $E=mc^2$—began to lift from behind the Earth, somewhere far beyond the Thames estuary; lifting, rolling, to embrace our lives.

I loved writing this book.

ISIS publish a wide range of books in large print, from fiction to biography. Any suggestions for books you would like to see in large print or audio are always welcome. Please send to the Editorial department at:

ISIS Publishing Ltd.
7 Centremead
Osney Mead
Oxford OX2 0ES
(01865) 250 333

A full list of titles is available free of charge from:
Ulverscroft large print books

(UK)
The Green
Bradgate Road, Anstey
Leicester LE7 7FU
Tel: (0116) 236 4325

(Australia)
P.O Box 953
Crows Nest
NSW 1585
Tel: (02) 9436 2622

(USA)
1881 Ridge Road
P.O Box 1230, West Seneca,
N.Y. 14224-1230
Tel: (716) 674 4270

(Canada)
P.O Box 80038
Burlington
Ontario L7L 6B1
Tel: (905) 637 8734

(New Zealand)
P.O Box 456
Feilding
Tel: (06) 323 6828

Details of **ISIS** complete and unabridged audio books are also available from these offices. Alternatively, contact your local library for details of their collection of **ISIS** large print and unabridged audio books.